EPP-Vol. 1

PACKAGING, RELIABILITY AND MANUFACTURING ISSUES ASSOCIATED WITH ELECTRONIC AND PHOTONIC PRODUCTS

presented at

THE 2001 ASME INTERNATIONAL MECHANICAL ENGINEERING CONGRESS AND EXPOSITION
NOVEMBER 11– 16, 2001
NEW YORK, NEW YORK

sponsored by

THE ELECTRONIC AND PHOTONIC PACKAGING DIVISION, ASME

edited by

I. CHARLES UME
GEORGIA INSTITUTE OF TECHNOLOGY

SURESH SITARAMAN
GEORGIA INSTITUTE OF TECHNOLOGY

JOHN LAU
AGILENT TECHNOLOGIES, INC.

K. RAMAKRISHNA
MOTOROLA, INC.

EPHRAIM SUHIR
IOLON, INC.

BAHGAT SAMMAKIA
BINGHAMPTON UNIVERSITY

GREG KOWALSKI
NORTHEASTERN UNIVERSITY

IMIN KAO
STATE UNIV. OF NEW YORK AT STONY BROOK

THE AMERICAN SOCIETY OF MECHANICAL ENGINEERS
Three Park Avenue / New York, N.Y. 10016

ISBN No. 0-7918-3565-0

Library of Congress Number 2001-134005

FOREWORD

Welcome to the 2001 ASME International Mechanical Engineering Congress and Exposition at the New York Hilton Hotel and Towers, New York, New York, USA. The Electronics and Photonics Packaging Division of the ASME invites you to attend the presentations of over 85 technical papers and panel discussions that span 23 technical sessions. These presentations will cover cutting edge research and innovations in such areas as: thermo-mechanics, photonics, nano- and micro-scale technology, reliability, silicon wafer, interconnections, thermal management, packaging, manufacturing, modeling and simulation, testing, MEMS and components as they relate to electronic and photonic packaging. The audience will include representatives from academia, government laboratories, and industries eager to stay abreast with the rapidly changing Global Electronics and Photonics Technology Evolutions/Revolutions.

The purpose of this conference is to continue the offering of high-quality coverage of technological innovations in the areas of design, manufacturing, process, materials, and testing for devices, components, and systems. For example, this year's program provides a look at emerging technologies in MEMS and nano- and micro-scale packaging. This conference offers those working in the field of electronic components, packaging, and physical design an excellent opportunity to stay current in the field.

This conference would not be possible without the hard work and dedication by the Symposium Organizers, and those authors who contributed their papers. I sincerely thank all of you, and I also thank the Session Chairs and Session Co-Chairs. My special thanks go to our expert panelists, who will be educating us on the new and future technological trends in the field.

I. Charles Ume
Program Chair
Georgia Institute of Technology
School of Mechanical Engineering
Atlanta, GA 30332-0405, USA
charles.ume@me.gatech.edu

CONTENTS

Proceedings of
2001 ASME International Mechanical Engineering Congress and Exposition
November 11–16, 2001, New York, NY
EPP-Vol. 1

IMECE2001/EPP-24700

WARPAGE ANALYSIS OF WAFERS WITH FILM COATING

Hai Ding, I. Charles Ume*, and Cheng Zhang
School of Mechanical Engineering
Georgia Institute of Technology
Atlanta, GA 30332
*Tel: (404) 894-7411
*Fax: (404) 894-9342
*Email: charles.ume@me.gatech.edu

ABSTRACT

Wafer-level packaging (WLP) is one of the trends of electronic packaging in the 21st century. Since 1994, many companies have released WLP licenses. One of the common concerns among these various approaches is wafer warpage. Warpage of wafer tends to introduces crack or delamination during dicing and low temperature storage process. After wafer dicing, warpage could reduce the quality of each package in the long run. Many documented works indicated that in the design and implementation of WLP, some key parameters have to be carefully considered and closely controlled to ensure higher packaging quality with the minimum warpage. For the case of wafer-level flip chip, the key parameters are Young's modulus, thickness, and coefficient of thermal expansion (CTE) of underfill.

In this research, an experimental design and statistical methods have been used to identify the model structure and parameters that are critical to the warpage of wafers. Regression models were identified based on the data obtained from finite element analysis (FEA) that is verified by shadow Moiré experiments. According to the models, the CTE, the coupling of Young's modulus and CTE, and the coupling of thickness and CTE of underfill primarily determine wafer warpage. Further FEA and shadow Moiré experiments indicate that the models are capable of predicting of wafer warpage in the process of WLP.

1 INTRODUCTION

High throughput is one of the main concerns under the continuous driving force of low cost in electronic packaging industry. WLP is a rapidly growing technology that gives solutions to lower cost. Many institutions and companies have invented various WLP technologies [1]. WLP is defined as the complete packaging of components at the wafer level. The die and package are manufactured, burned in, and tested on the wafer prior to singulation. The batch process gives the advantage of lower cost per chip. This is more obvious for wafer-level flip-chip process [2]. Eliminating individual flux application, underfill dispensing, and burn-in & testing procedures can save a great amount of time and capital.

One challenge of WLP is wafer warpage. Figure 1 shows a wafer-level flip-chip process flow [2-3]. Due to the huge aspect ratio of wafer--usually with diameter to thickness around 200:1, and CTE mismatch between film coating and wafer--around 10:1 or larger, out of thickness warpage becomes prominent. The problem is aggravated when film of underfill is dispensed at an elevated temperature, while stored at a lower temperature, e.g., room temperature. Warpage is a global effect that reflects inter-layer residue stresses, which could introduce die crack, fatigue, or delamination between wafer and film. Also, warpage affects subsequent dicing and individual packaging process.

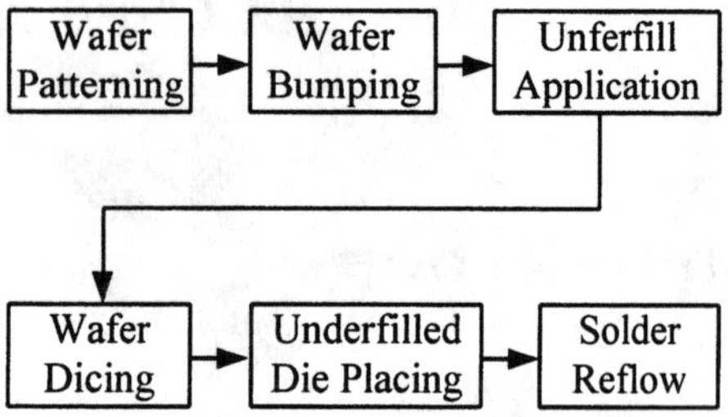

Figure 1. Wafer-level flip-chip process.

In this paper, wafer-level flip-chip configuration was taken as an example and warpage of wafers with underfill was studied. In section two, a finite element (FE) model was built. A design of experiment aiming to extract individual and interacting effects among Young's modulus, thickness, and CTE of underfill was constructed. Warpage was calculated from the FE model and the results were validated by shadow Moiré experiment, which is described in section three. In section four, analysis of variance (ANOVA) was performed to identify effects of undefill properties on wafer warpage. Consequently, regression models were constructed and shown to give satisfactory estimation of warpage. In the end, a conclusion was drawn.

2 FINITE ELEMENT ANALYSIS

Finite element analysis (FEA) saves time and material for preliminary warpage study. Wafers and film materials are expensive. The process of film coating is complex and requires special equipment. Once validated by shadow Moiré experiment, FEA is the viable way to calculate warpage. It is especially less expensive when doing repetitive tasks with only a few parameters changing. Critical issues of FEA are geometry modeling, material modeling, finite element selection, meshing, and loading application.

In this research, the FE model is simplified to emphasize warpage analysis methodology, in which FEA and experiments are means of getting accurate warpage data. Microelectronic manufacturing involves complex geometry, materials, and processes. The FE model can always be refined to represent each specific application. However, geometry with two layers is presented here with constant material properties. In addition, since only the effect of underfill material on the warpage is concerned, the initial warpage of bumped wafer is taken to be zero.

2.1 Geometry Modeling

A typical 150-mm diameter (6-inch) wafer has a thickness of 700 μm. The film could have a thickness of 100~200 μm. Figure 2 shows the cross section of a wafer with film coating. As can be seen, this geometry bears a huge aspect ratio between diameter direction and through thickness direction. This is one reason why warpage becomes prominent. Also this makes finite element meshing using solid element more difficult than using shell element or plane element.

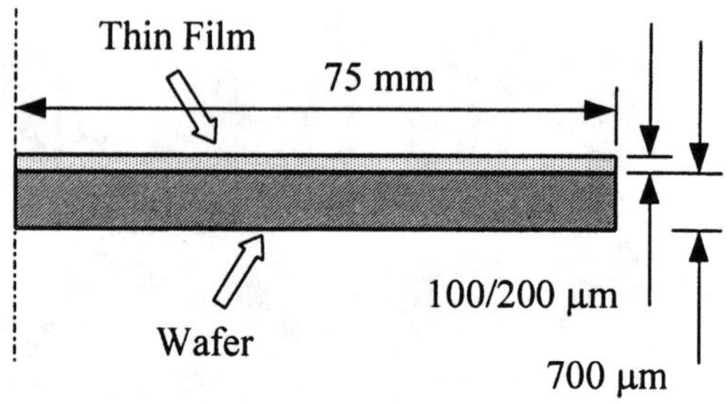

Figure 2. Cross section of a wafer with film.

The structure shown in Fig. 2 could be a generic underfilled wafer in warpage analysis by taking the following assumptions:

- Various deposition and growth layers (epitaxy, oxidation, metallization, passivation, etc.) are so thin, in the order of several microns, that they can be neglected.
- When solder bumps are sparce, they are tiny compared to wafer and underfill layers and can be neglected. These small structures only affect stress, strain, and displacement locally. For the case of dense solder bumps, introducing a uniform layer incorprating effective solder properties can better represent the structure.
- Orientation flat (a small cutoff along the chord of a wafer to label its crystal orientation) affects stress, strain, and displacement only locally, so they can be neglected.

Warpage is an accumulated effect across speciman. Therefore, neglecting subtle structures is acceptable. Nevertheless, one can always refine and detail the model for a specific application. From the above assumptions, the model can be simplified to 2D as shown in Fig. 2. But to better visualize the shape of the warpage, 3D solid geometry was actually used.

2.2 Material Properties

Silicon wafer has Young's modulus of 130 GPa, Poisson's ratio of 0.28, and CTE of 3 ppm/K. There is a large CTE mismatch between silicon wafer and underfill, as shown in Table 1. Typical underfill without filler has a CTE of 80 ppm/K.

Underfill with filler has a larger Young's modulus and smaller CTE than underfill without filler.

	E (GPa)	ν	CTE (ppm/K)
Si wafer	130	0.28	3
Underfill w/o filler	3	0.30	80
Underfill w/ filler	10	0.30	25

Table 1. Material properties of silicon wafer and underfill.

Note that most microelectronic materials bear temperature and/or time dependent properties. In this study, the temperature (25-100 °C) and time span is relatively small. Therefore, the material properties are assumed constant.

2.3 Element Selection

Three types of element were tried out in this research. They are 2D element, shell element, and 3D element [4]. 2D and 3D element all have mid-node. Shell element used is an element that can hold up to 16 structure layers.

Generally speaking, solid element is the most accurate, although it is more expansive in the sense of meshing and computation time [5]. Also, element with mid-node will give more accurate results. But for this simple problem, we can expect no difference from all of these elements.

2.4 Meshing and Loading

2D mesh is the simplest and cheapest. 3D mesh is the most expensive. While shell mesh is much more efficient for shell-like structures. All the above three meshes were constructed to have comparable element sizes.

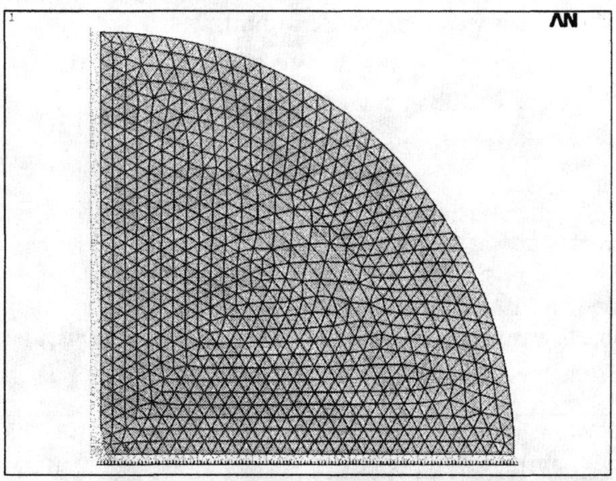

Figure 3. Mesh by using solid element.

Mesh of a quarter of the wafer by using solid element is shown in Fig. 3. The mesh also has been doubled to conduct convergence study. From symmetry, the left side is constrained in horizontal direction and the bottom side is constrained in vertical direction. The bottom-left corner is constrained in all degree of freedom (DOF) and is taken as the reference of warpage.

The loading in this analysis is temperature. A uniform temperature of 25 °C is applied to the whole structure, with 100 °C as the reference temperature. This is the scenario when underfill is cured at 100°C and is stored at room temperature.

2.5 Design of Experiment

To fully capture the information of how different factors affect warpage, a three-factor, two-level, and full-factorial design of experiment was conducted. As shown in Table 2, the three factors are Young's modulus, CTE, and thickness of underfill. Each factor has two levels. Totally there are eight runs. Although Young's modulus and CTE are not independent to each other, as shown in Table 1, the virtual design was adopted to get an insight of the influence of each parameter.

No. of Runs	V1 (E, Gpa)	V2 (CTE, ppm/K)	V3 (Thick., mm)
1	3	25	0.1
2	10	25	0.1
3	3	80	0.1
4	10	80	0.1
5	3	25	0.2
6	10	25	0.2
7	3	80	0.2
8	10	80	0.2

Table 2. 3-factor, 2-level, and full-factorial DOE.

2.6 FEA Results

As shown in Table 3, results of Run 2 using different elements are very close to each other for this simple problem. When the mesh is doubled, the change is barely noticeable if not zero. Therefore, it is confirmed that the developed FE model is consistent and convergent. Figure 4 shows a deformed mesh along with the undeformed line frame.

Configurations	Results (mm)
Plane element, half symmetry w/o flat	0.48650
Shell element, half symmetry w/ flat	0.48665
Shell element, quarter symmetry w/o flat	0.48666
Solid element, quarter symmetry w/o flat	0.48652

Table 3. Comparison of results using different finite elements of Run 2.

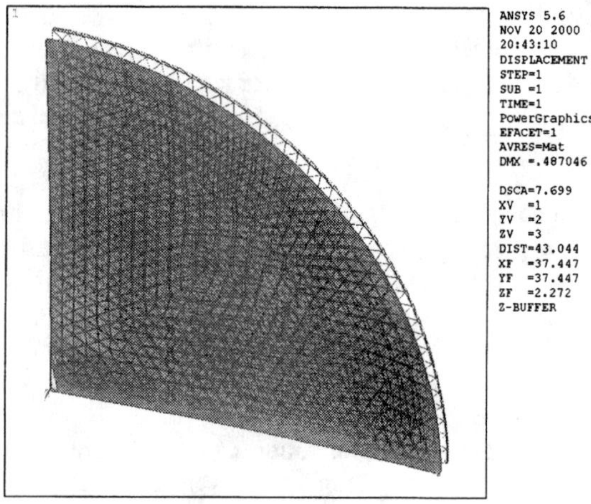

Figure 4. Deformed shape along with undeformed line frame of Run 2.

3 SHADOW MOIRÉ EXPERIMENTS

To validate the results from FEA, experiments using shadow Moiré technique [6] were conducted. Fig. 5 shows a typical setup of shadow Moiré experiment. As can be seen, white light projects the shadow of the grating onto the sample surface. The deformed grating shadow interferes with the grating itself to form Moiré pattern. The CCD camera captures the Moiré image and sends it to computer to calculate warpage. Fig. 6 shows the Moiré fringe under condition of Run 2 and the calculated surface plot. Note that the surface plot is a rectangular crop out of a circular wafer.

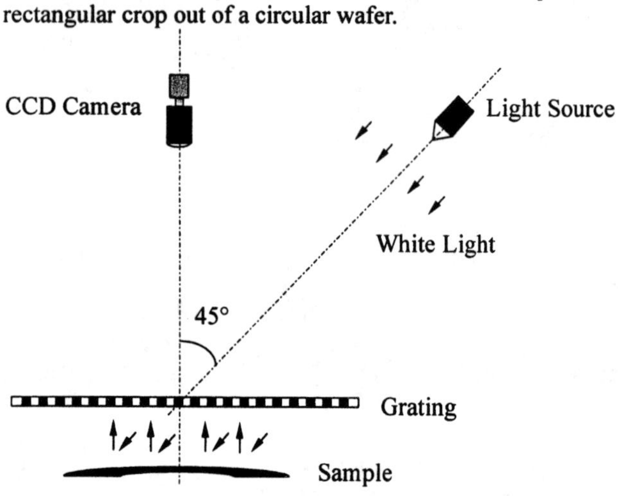

Figure 5. A typical shadow Moiré setup.

Relatively good match between FEA and shadow Moiré experiments was achieved. The 8% error between FEA (0.48652 mm) and shadow Moiré experiment (0.53086 mm) could come from non-uniform coating process, inaccurate material properties, or unaccounted creep and plasticity of underfill. In addition, chemical shrinkage could add to higher warpage values than predicted by FE model. This validation of FE model by shadow Moiré experiments paves the way of the following statistical analysis.

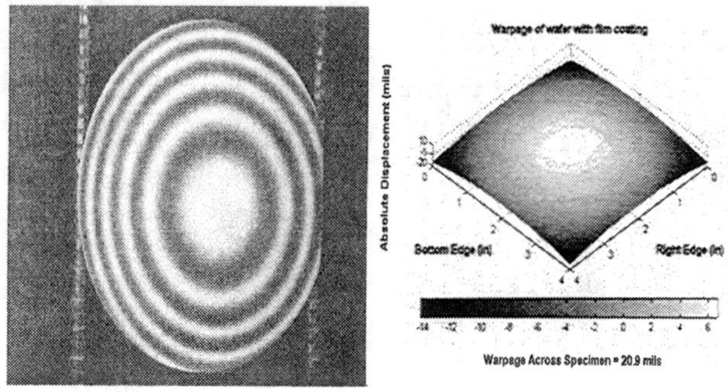

Figure 6. Fringe pattern and surface plot of shadow Moiré result of Run 2.

4 MODEL IDENTIFICATION AND VERIFICATION

4.1 Variable Effects

Results of research and observation of laboratory experiments indicate that warpage of wafers is affected by a variety of factors, such as Young's Modulus of underfill and wafer, coefficient of thermal expansions (CTE) of underfill and wafer, thickness of underfill and wafer, temperature profile, moisture profile, etc. These system independent variables and the interacting effects of these variables have their influence on warpage of wafers. In this study, individual and interacting effects of three independent variables, namely the Young's Modulus of underfill, the CTE of underfill, and the thickness of underfill, are illustrated by Eq. (1).

$$y = y(x_1, x_2, x_3) \tag{1}$$

where,

y is warpage of wafers (mm),

x_1 is Young's Modulus E of underfill (GPa),

x_2 is CTE of underfill (ppm/K),

x_3 is thickness of underfill (mm).

A three-factor, two-level, and full-factorial design of experiment was conducted, as shown in Table 2. The replicate of each run is one. Table 4 shows the warpage data of wafers based on FEA results. The data of warpage of the second run in Table 2 was validated by shadow Moiré method, as discussed in section 3.

Warpage (mm)	X1-Young's E (GPa)	X2-CTE (ppm/K)	X3-Thick. (mm)
0.15146	3	25	0.1
0.48652	10	25	0.1
0.53049	3	80	0.1
1.7032	10	80	0.1

0.33302	3	25	0.2
1.0170	10	25	0.2
1.1660	3	80	0.2
3.5599	10	80	0.2

Table 4. Data of warpage—FEA results.

The technique of analysis of variance (ANOVA, Balanced Designs) was conducted to investigate the main effects from the three independent variables based on the FEA warpage data. Following is the output of ANOVA.

Source	DF	SS	MS	F	P
X1-Young's E	1	2.6285	2.6285	6.53	0.063
X2-CTE	1	3.0896	630896	7.67	0.050
X3-Thickness	1	1.2834	1.28347	3.19	0.149
Error	4	1.6107	0.4027		
Total	7	8.6122			

Table 5. Output of ANOVA.

It is shown from F-distributions and P-values in Table 5 that if the one-sided confidence interval is set as 95%, only X2 (CTE) has significant effect on warpage while there is no evidence of significant effects resulting from X1 (Young's Modulus) and X3 (thickness). If the one-sided confidence interval is set as 90%, both X1 (Young's Modulus) and X2 (CTE) have significant effects on warpage while X3 (thickness) has no evidence to confirm the significant effect on warpage. Therefore, the conclusion of this ANOVA output is that among these three factors, CTE has the most significant effect on warpage while thickness has the least effect. In the following subsection, multivariable regression analysis is used to determine models capable of explaining the variability of the warpage data.

4.2 Regression Analysis for Model Determination and Predictability Testing

Equation (1) shows that warpage is a function of Young's Modulus, CTE, thickness of underfill, and their interaction terms. Therefore, a full bilinear model can be given as:

$$y=\beta_0+\beta_1x_1+\beta_2x_2+\beta_3x_3+\beta_4x_1x_2+\beta_5x_1x_3+\beta_6x_2x_3 \quad (2)$$

Typically, not all the terms give useful explanations to experimental data. It is beneficial to identify a reduced model relative to the given model, when a reduced model provides an adequate explanation of the experimental data. In fact, for data collected in a factorial experimental design, a reduced model often provides better prediction and parameter estimation than a full model. Generally, there are three fundamental statistics that are most commonly used in best subset model selections, namely R^2, Adj R^2, and Mallows' C_p [7]. The term R^2 statistically measures the percentage of the total variability in the dependent variable that is explained by a regression model. It is defined as

$$R^2 = \frac{SSR}{SST} \quad (3)$$

where,

SSR is the sum of squared regression error,
SST is the sum of squared total error.

The criterion of R^2 has a major drawback, as its value will increase with the number of variables involved in the model. This characteristic makes it difficult in comparing different models when they have unequal numbers of predictor variables. The Adjusted R^2 is simply an adjustment of R^2 that allows comparisons to be made between models with different numbers of predictor variables. Adjusted R^2 is given as

$$\text{Adj } R^2 = 1 - \frac{n-1}{n-r}(1-R^2) \quad (4)$$

where,

n is the number of data points.
r is the number of predictor variables.
R^2 is the coefficient of multiple determination as defined in Eq. (3).

Based on the Adjusted R^2 criterion, the model with largest Adjusted R^2 is usually the best model. Mallows's C_p statistic estimates a measure of the difference between the fitted regression surface from a reduced model and the actual regression surface. The C_p criterion says that the best-reduced model is associated with the smallest C_p value and it is close to the number of predictor variables. Mallows' C_p is given as

$$C_p = \frac{SSE(R)}{MSE(F)} - (n-2r) \quad (5)$$

where,

SSE(R) is the sum of squared error of a reduced model,
MSE(F) is the mean squared error of a full model,
n is the number of data points,
r is the number of predictor variables.

Table 6 gives the extended data of warpage applied for the best subset regression analysis and regression model determination.

Warpage (mm)	X1	X2	X3	X1X2	X1X3	X2X3
0.15146	3	25	0.1	75	0.3	2.5
0.48652	10	25	0.1	250	1	2.5
0.53049	3	80	0.1	240	0.3	8
1.7032	10	80	0.1	800	1	8
0.33302	3	25	0.2	75	0.6	5
1.0170	10	25	0.2	250	2	5
1.1660	3	80	0.2	240	0.6	16

3.5599	10	80	0.2	800	2	16

Table 6. Extended data for regression analysis.

Table 7 shows the statistics output based on the full model by using the data in Table 6. The best-reduced model with X2, X1X2, and X2X3 has the smallest Mallows's C_p of 4.3 with larger Adjusted R^2 of 91.7%, and the larger R^2 of 95.3%. It is the best subset of the given full model in Eq. (2) based on all criteria. Then, the multivariable regression model of the reduced model is given in Eq. (6).

Vars	R^2	R^2 (adj)	C_p	$\sqrt{MSE}$	X1	X2	X3	X1X2	X1X3	X2X3
1	75.6	71.6	18.1	0.59155				X		
1	55.4	47.9	36.4	0.80050						X
2	90.5	86.7	6.6	0.40402			X	X		
2	86.8	81.5	10.0	0.47755				X		X
3	95.3	91.7	4.3	0.31873		X		X		X
3	94.6	90.5	4.9	0.34238				X	X	X
4	96.0	90.6	5.7	0.34049		X		X	X	X
4	95.9	90.3	5.7	0.34464	X			X	X	X
5	97.2	90.1	6.6	0.34964	X	X		X	X	X
5	96.2	86.9	7.4	0.40191		X	X	X	X	X
6	98.9	92.3	7.0	0.30839	X	X	X	X	X	X

Table 7. Best subsets regression

$$\hat{Y} = -0.068 - 0.0211x_2 + 0.00316x_1x_2 + 0.155x_2x_3 \quad (6)$$

where, $\hat{Y}$ is the fitted value of warpage (mm).

Equation (6) indicates that the CTE of underfill, the coupling of CTE and the Yong's Modulus of underfill, as well as the coupling of CTE and Thickness of underfill have dominant influence on warpage. With this optimal reduced model, parameters that estimate warpages are allowed to be assessed under a variety of operation conditions without prior calibration.

From Table 7, however, the full regression model has the largest Adjusted R^2 of 92.3%, and the largest R^2 of 98.9% with the Mallows's C_p of 7.0 which is close to the number of variables–6 used for this regression analysis. Therefore, the full model is also a possible good choice for the model determination. Thus, a full multivariable regression model could be given in Eq. (7),

$$\hat{Y} = 1.16 - 0.178x_1 - 0.0232x_2 - 7.78x_3 + 0.00331x_1x_2 + 1.12x_1x_3 + 0.162x_2x_3 \quad (7)$$

Figure 7 shows the normal probability plot of the residuals for the reduced model. While, Figure 8 shows the residuals versus the order of the data for the reduced model. From the diagnostic figures, it is known that the pattern of normal probability plot of the residuals of the reduced model shows a linear tendency, and the pattern of residuals versus the order of the data looks random. It is implied that all the model assumptions and analysis used for this determination of the best reduced model are correct.

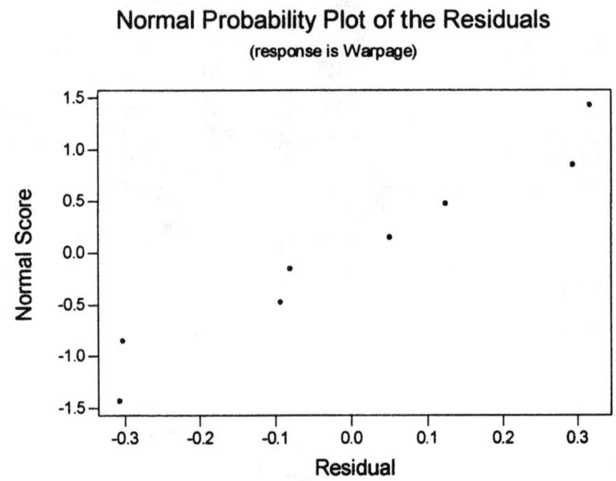

Figure 7. NPP of the reduced model.

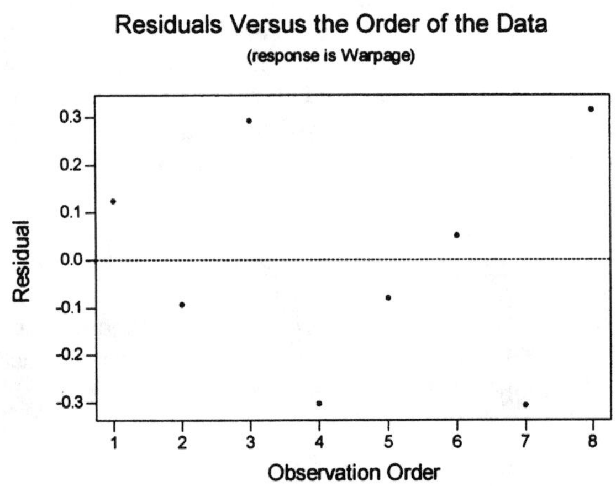

Figure 8. Residuals vs. order of the reduced model.

Based on the warpage data obtained from another two independent tests by FEA, the developed models were verified by both interpolation and extrapolation. When setting the variables within the range of factor levels, such as, E=6 GPa, CTE=55 ppm/K, and thickness=0.15 mm, the warpage of the wafer is 1.1000 mm. When setting the variables outside the range of factor levels, such as, E=12 GPa, CTE=96 ppm/K, and thickness=0.24 mm (all 20% higher), the warpage of the wafer is 6.0614 mm. Table 8 shows the results of the residual analysis and corresponding estimation errors.

	Interpolation		Extrapolation	
	Residual (mm)	Relative Error	Residual (mm)	Relative Error
Full Model	0.0142	1.29%	0.3606	5.95%
Reduced Model	0.0070	0.64%	0.9434	15.56%

Table 8. Residuals and estimation errors.

Based on the above table, when the variables are located within the calibrated range of the factor levels, the estimation error is 1.29% and 0.64% for full and reduced model respectively. When the variables (20% increment) are located outside the range of the factor levels that have already been calibrated, the full regression model still provides good estimation with relative error around 6% while the reduced model only gives rough estimation with 15.56% relative error. It is shown clearly that both models developed in this study have the capability to accurately estimate the warpage of wafer within the ranges of the factors.

5 CONCLUSIONS

This study analyzed the effects of the Young' Modulus of underfill, the coefficient of thermal expansion (CTE) of underfill, and the thickness of underfill on warpage of wafers during the packaging process. A full-factorial experimental design and the statistical regression analysis method have been used to identify the model structure and parameters that best estimate wafer warpage. Based on the data obtained from FEA that is also verified by shadow Moiré method, warpage of wafers is mainly determined by the CTE, the coupling of Young's Modulus and CTE, and the coupling of thickness and CTE. A reduced and a full regression model were identified based on statistical theory. Models were then verified by evaluating independent factor levels under untested conditions to test the predictability. The experimental results (FEA data) indicate that the models developed in this research have the capability to provide accurate estimations of warpage of wafers during the packaging process.

ACKNOWLEDGMENTS

The authors would like to thank Dr. Jicun Lu of Packaging Research Center of Georgia Institute of Technology for preparation of specimen and providing material properties. They also appreciate Reinhard Powell's participation when doing initial part of this work as a class project.

REFERENCES

[1] Garrou, P., 2000, "Wafer Level Chip Scale Packaging (WL-CSP): An Overview" *IEEE Transactions on Advanced Packaging*, 23, no. 2, May, pp. 198-205.

[2] Johnson, C.D., Busch, S.C., and Baldwin, D.F., 1999, "Wafer Scale Packaging Based on Underfill Applied at the Wafer Level for Low Cost Flip Chip Processing," Proceedings of SPIE, vol. 3830, pp. 371-375.

[3] Shi, S., Yamashita, T., and Wong, C.P., 1999, "Development of the Wafer Level Compressive-Flow Underfill Process and Its Involved Materials," *IEEE Transactions on Electronics Packaging Manufacturing*, 22, no. 4, October, pp. 274-281.

[4] ANSYS, Inc., 1998, "ANSYS Elements Reference, Tenth Edition."

[5] Variyam, M.N., Xie, W., and Sitaraman, S.K., 2000, "Role of Out-of-Plane Coefficient of Thermal Expansion in Electronic Packaging Modeling", *ASME Journal of Electronic Packaging*, 122, no. 2, pp. 121-127.

[6] Stiteler, M., Ume, I.C., 1997, "System for Real-Time Measurements of Thermally Induced Warpage in a Simulated Infrared Soldering Environment", ASME Journal of Electronic Packaging, 119, pp. 1-7.

[7] Christensen, R., 1996, "Analysis of Variable, Design and Regression, Applied Statistical Methods", Chapman & Hall, pp. 414-431.

Proceedings of
2001 ASME International Mechanical Engineering Congress and Exposition
November 11–16, 2001, New York, NY
EPP-Vol. 1

IMECE2001/EPP-24701

AN ACTIVE RADIAL COUNTERCURRENT HEAT SINK DRIVEN BY A SYNTHETIC JET ACTUATOR

Jelena Vukasinovic[†]
Woodruff School of Mechanical Engineering
Georgia Institute of Technology
Atlanta, GA 30332
email: gt3133a@prism.gatech.edu

Ari Glezer
Woodruff School of Mechanical Engineering
Georgia Institute of Technology
Atlanta, GA 30332
email: ari.glezer@me.gatech.edu

ABSTRACT

The performance of a low-profile radial countercurrent heat sink driven by an integrated synthetic jet actuator is investigated experimentally. A packaged thermal test die is cooled using an array of synthetic jets normally impinging on the extended surface. A power dissipation of 50 W is accomplished at the nominal case temperature of $T_c = 70$ °C.

The heat sink design is driven by the flow and heat transfer analysis of normal jet impingement in a confined flow geometry consisting of two parallel circular plates having a diameter that is typically an order of magnitude larger than the spacing between the plates. The velocity and temperature distributions are measured using particle image velocimetry and arrays of thermocouple sensors. A jet actuator is integrated into one of the plates and cools a test heater attached to the opposite surface. The jet draws its makeup air from ambient, impinges on the heater, and ultimately rejects the heat to ambient. This introduces a radial countercurrent flow in the gap between the plates that includes a layer of hot air dispensed along the top plate and a layer of cooler ambient air entrained along the jet exit plane. When the jet is activated the heater temperature drops substantially. Although the global heat transfer coefficient decreases with decreasing gap height, flow pathlines show that the jet can still entrain cool air from ambient and effect substantial cooling even when the spacing between the plates is of the order of the jet orifice diameter.

[†] Address all correspondence to this author

I. INTRODUCTION

The increasing demand for efficient cooling of integrated circuits within small, confined housings has motivated the development of novel fluidic heat removal devices that can be integrated with dense, high-performance electronics and photonics. Earlier work has demonstrated the utility of forced convection cooling using small-scale synthetic jet actuators (Mahalingam and Glezer, 2001) and has suggested that this technology can be adapted to complex housing geometries and confined spaces.

Commercial air-cooling of high-speed microprocessors typically utilizes forced convection induced by a fan. Although fans generally provide high volume flow rates over extended surfaces, they may be inefficient in terms of heat dissipation per volume flow rate because of the weak thermal coupling between the cooling air and the surface. Such coupling may be achieved by the impingement of conventional-steady or pulsed jets onto the surface, but the implementation of this cooling approach is often hindered by complicated plumbing and fluidic packaging and the fact that miniaturization leads to a reduction in jet Reynolds number and therefore the cooling efficiency. Recent work has demonstrated that synthetic jets induce small-scale motion near solid surfaces and provide efficient mixing with global cross flow (Russell, 1999). An important feature of synthetic jets is that while they are zero net mass flux in nature, they have non-zero streamwise momentum that eliminates the need for complex plumbing required for conventional jets. As shown by Smith and Glezer (1998) these jets are synthesized from entrained ambient air owing to the motion of a diaphragm that is mounted onto one of the walls of a sealed shallow cavity.

Although there is an extensive body of work on continuous (Jambunathan et al., 1992) and pulsed (Azevedo et al., 1994) jet cooling, the air-cooling using synthetic jets is relatively new. Thompson et al. (1997) investigated an axisymmetric synthetic jet impingement to cool a 49-element MOSFET mutli-chip module and reported 250% increase in heat dissipation over natural convection at a die temperature of 100 °C. More recently, Mahalingam and Glezer (2001) integrated a synthetic jet actuator with a conventional heat sink and by exploiting the jet flow in the channels between the fins, achieved heat removal of 110 W at 100 °C in a package having a volume of 200 cm³. While the work of Mahalingam and Glezer has primarily focused on the development of active heat sinks integrated with synthetic jets for relatively high-power integrated circuits (e.g., microprocessors), the present work focuses on fluidic-based thermal management that specifically targets heat transfer in confined geometries that would otherwise be difficult to cool using conventional global approaches. This work investigates the heat transfer and fluid dynamics of a normally impinging synthetic jet localized over the test heater in a confined flow configuration. Section II describes the experimental setup and discusses the relevant flow and heat transfer parameters. Section III describes the dynamics of the countercurrent flow that develops in the narrow gap between the parallel plates by a centrally impinging synthetic jet and presents guidance to prevent vortex reingestion and recirculation to maximize heat removal. Section IV focuses on the heat transfer measurements for the baseline geometry. Section V describes the performance of an active radial countercurrent heat sink prototype that utilizes the baseline geometry and incorporates triangular fins between the parallel plates to enlarge the heat transfer surface area. Within the heat sink, the synthetic jet impingement induces periodic thinning of the thermal boundary layer and introduces countercurrent radial flow through the heat sink channels to entrain ambient air and transfer heat to the ambient. Section VI presents conclusions from this work.

NOMENCLATURE

d	Synthetic jet orifice diameter
d_h	Test-heater diameter
D	Parallel plates diameter
h	Heat transfer coefficient
H	Gap height between the parallel plates
L	Jet stroke length $L = \int_0^{\tau/2} u_o(t)dt$
R_e	Jet Reynolds number $R_e = Ud/\nu$
t	Time
T	Temperature
ΔT_{ca}	Case to ambient temperature difference
$u_o(t)$	Centerline velocity in jet orifice
U	Averaged jet velocity $U = L/\tau$
ν	Kinematic viscosity
θ_{ca}	Case to ambient thermal resistance
τ	Period of the jet cycle

II. EXPERIMENTAL SETUP

The test setup (Figure 1), consists of two parallel circular plates each having a diameter, D = 152.4 mm, that is typically much larger than the adjustable gap height between the plates, H. A jet actuator is integrated into the bottom plate and a kapton test-heater ($d_h/d = 5.3$) is placed on the center of the top surface. The actuator consists of the plenum and an electromagnetically driven diaphragm located at the bottom of a shallow cavity with a 6.35 mm orifice in the bottom plate. Top surface is insulated to minimize conduction losses through its back and contains arrays of T-type thermocouples integrated to both sides of the plate. The thermocouple readings are digitized using a 12-bit A/D board having a cold-junction compensation with quantization errors of less than 0.05 °C.

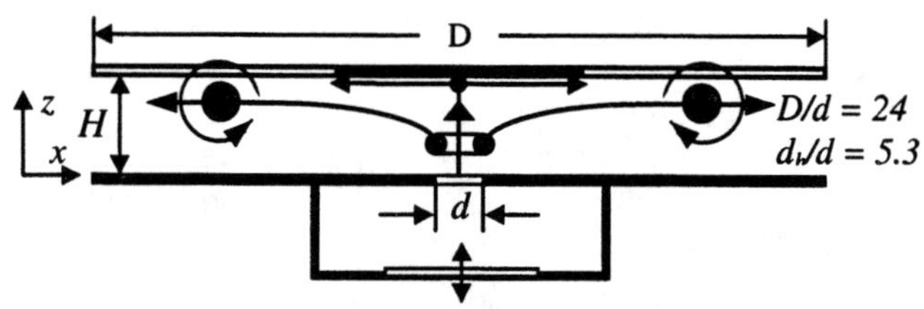

Figure 1. Schematic of the setup

As argued by Smith and Glezer (1998), the synthetic jet may be characterized by two primary dimensionless parameters: the dimensionless stroke length, L/d, and the jet Reynolds number. The stroke length, L, measures the displacement of the slug of fluid pushed from the orifice during the blowing stroke ($\tau/2$), while the jet Reynolds number, R_e, is based on the time-averaged jet orifice velocity during blowing stroke, U. In the present work, it is important to comment on the effects of the gap height to plate diameter ratio, H/D, to ensure that the jet is capable to entrain cooler ambient air and to prevent the recirculation of hot air. For a fixed jet exit velocity, $H/D \ll 1$ enhances flow separation during blowing stroke, following the jet impingement on the heated surface, and may lead to recirculation. Moreover, a decrease in H/D increases the losses associated with ambient air entrainment. In contrast, the increase in dimensionless stroke length, L/d, leads to a greater radial dispalcement of dispensed air, with larger jet Reynolds number thus decreasing the potential for the formation of recirculating flow domains within the gap and increasing the stagnation heat transfer.

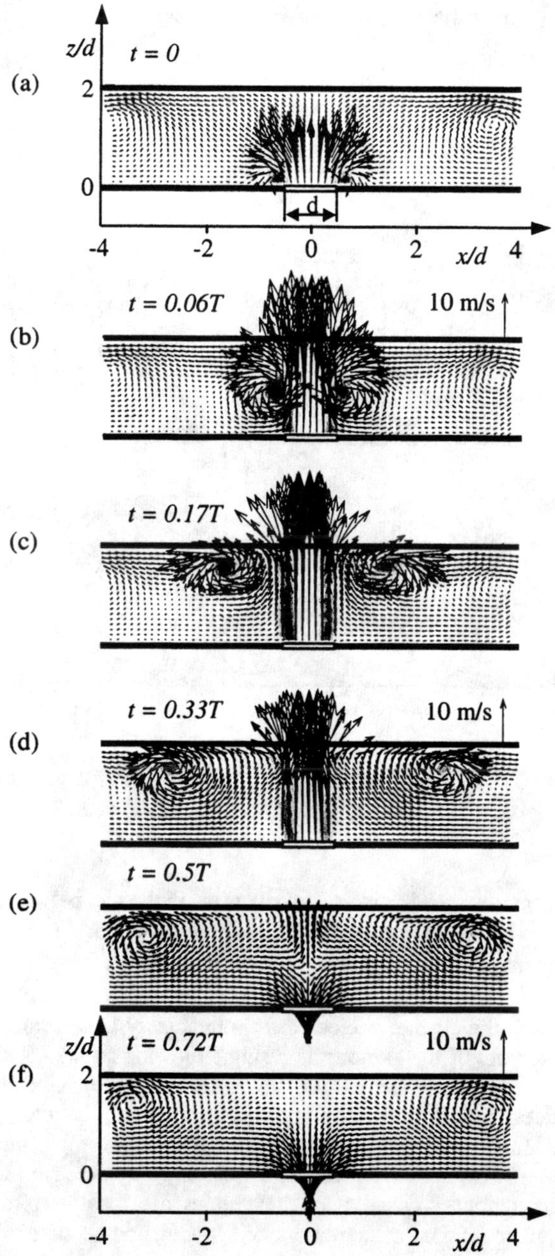

Figure 2. Velocity field in the vertical plane $H/D = 0.08$

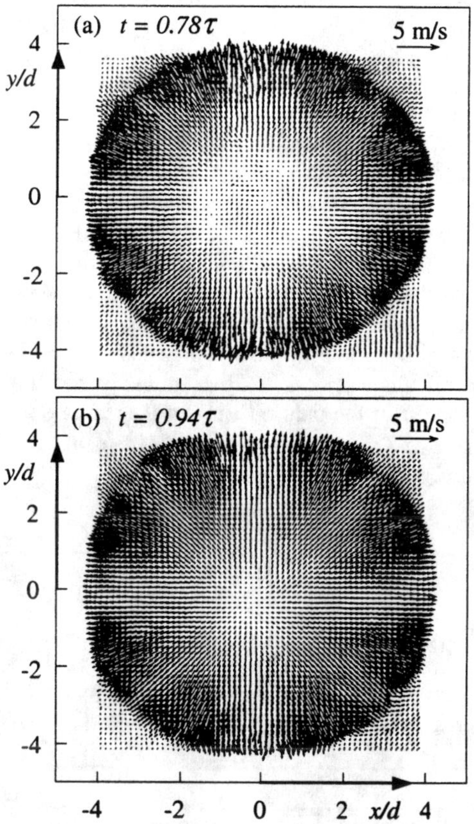

Figure 3. Flow near the impingement surface during suction stroke, $H/D = 0.08$

III. FLOW DYNAMICS

The flow field between the plates in a vertical plane that includes the jet centerline is measured using particle image velocimetry (PIV). The measurements are taken within the closed glass enclosure using smoke as seed particles. A sheet of light from a double-pulsed Nd-YAG laser is used to illuminate the flow phase-locked to the driving signal. Image pairs are captured using a 1008x1016 pixel CCD camera. The system magnification is such that the nominal particle size in the image is sub-pixel. Velocity vectors are computed using standard cross-correlation algorithms. The nominal velocity resolution is 1% of the peak velocity. Images of the phase-averaged flow at six instances during the excitation cycle (80 Hz), averaged over 100 realizations, are shown in Figures 2(a-f), within the gap for which the ratio of gap height to plate diameter is $H/D = 0.08$. The measurement domain ($2d \times 8d$) is bounded by the orifice plate at the bottom and the top plate. The peak orifice velocity of the free-jet (measured using hot wire anemometry) is 22 m/s. This yields the jet Reynolds number of $R_e = 3300$ and its dimensionless stroke length of $L/d = 15$. The oscillations of the membrane cause cyclic ingestion and ejection of fluid into and from the orifice, bounding the cavity. The ejection induces a vortex ring at the sharp edge of the orifice, due to flow separation, which moves away from the orifice under its own momentum, Figures 2(a-b). Upon impingement on the top plate, Figure 2c, vortex ring stretches radially outward pushing the hot air near the top surface toward the radial edge of the plate,

Figures 2(d-f). It is noted that in Figures 2(b-e) air does not penetrate the top surface, these vectors correspond to the magnitudes of velocity in the vicinity of the top plate. During the suction stroke, air is entrained into the cavity along the bottom plate, Figures 2(e-f).

The flow in the horizonatal plane near the impingement surface, $z/H = 0.9$, is also studied using PIV. Phase-averaged images show that the flow is directed from the stagnation point radially outward throughout the entire cycle. Figures 3(a-b) indicate this for two instances during the suction stroke. In these images the vortex ring is out of the field of view and no radial inflow, towards the jet centerline, is detected. This suggests that the ejected fluid is not ingested into the jet cavity during the following suction stroke. It should be noted however, that as the speed of the induced wall jet decreases downstream from the orifice ($x/d > 5$), some of the dispensed air might subsequently recirculate back into the jet cavity.

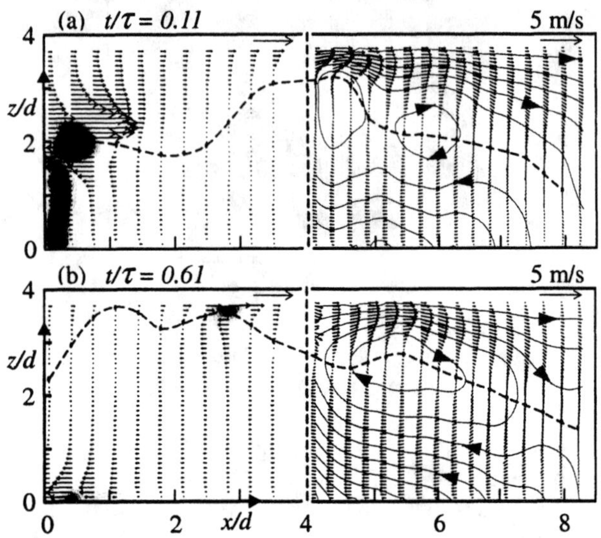

Figure 4. Radial countercurrent flow during (a) blowing and (b) suction stroke for $H/D = 0.17$

The time-periodic jet formation induces a radial countercurrent flow within the gap. Figures 4(a-b) show the radial countercurrent flow for two phase points, one during blowing and one during suction stroke, in the gap having an H/D of 0.17. In these images, the jet centerline coincides with the left side of the image and the measurement domain spreads radially to more than $8d$. This flow field includes a layer of heated air that is expelled along the top surface, and a layer of entrained, cooler ambient air along the bottom surface. The dashed line in these images represents a cluster of points for which the magnitude of radial velocity is zero that separates these two flow regions. In Figure 4a the flood contour points to vortex location while the computed streamlines, marked by arrows, show the formation of instantaneous recirculating flow domains.

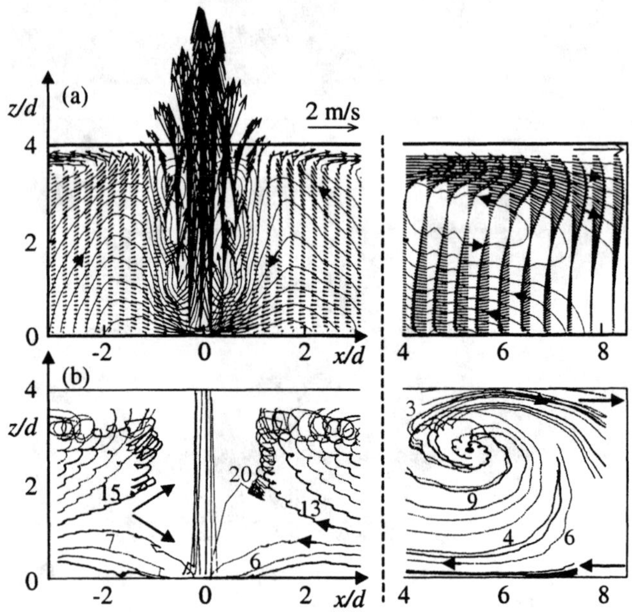

Figure 5. (a) Time-averaged velocity distributions in the gap and (b) pathlines for $H/D = 0.17$

Time-averaged velocity distributions in the gap between the parallel plates are shown in Figure 5(a) for $H/D = 0.17$. Closed streamline contours indicate the fluid recirulation at $x/d \approx 5.5$. When the gap aspect ratio is small, as in the present geometry, the air that is ejected from the cavity is generally not advected out of the gap during a single blowing stroke. Rather, it takes a finite number of jet strokes for the air to sequentially advances radially outward and leaves the gap between the plates. In Figure 5b, the pathlines, computed from the phase-averaged flow assuming that the flow is axisymmetric, show this sequential flow advance. Pathlines are locally spiraling along their primary direction due to blowing and suction over a diaphragm cycle. Every pitch of these irregular spirals therefore corresponds to one cycle of the driving signal. The number of cycles that are needed for a given particle to propagate along its pathline is denoted next to each pathline. During the suction stroke, the entrained air is partly ingested into the cavity and partly entrained by the vortex itself during its upward axial motion. The pathlines in Figure 5b demonstrate this entrainment into the vortex. The recirculation domain noted in Figure 5a appears to have a significant transfer of fluid across its boundary. Pathlines show that fluid enters and leaves this region over a finite number of jet cycles. Despite the recirculation, however, the radial countercurrent flow is present along the full length of the gap.

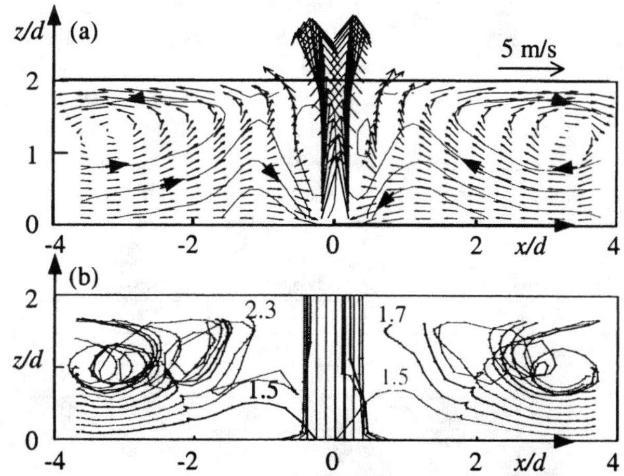

Figure 6. (a) Time-averaged velocity distributions in the gap and (b) pathlines for $H/D = 0.08$

There are two competing effects that affect the flow when the spacing between the plates is reduced. The losses for ambient air entrainment increase, but the volume fraction of expelled fluid relative to the volume of fluid in the gap increases. To illustrate this, Figures 6(a-b) show the time-averaged velocity distributions and flow pathlines in the gap having an H/D of 0.08. These measurements demonstrate higher velocity and shorter particle residence times near the orifice compared to the case of $H/D = 0.17$. Therefore for $H/D = 0.08$ the losses for entrainment appear to be reduced by the increase in magnitude of velocity in the gap. Other measurements (not shown) have indicated that a further decrease in H/D leads to the blockage of ambient air entrainment owing to an increase in the radial pressure losses.

As reported by Smith et al. (1999), an axisymmetric synthetic jet having a dimensionless stroke length of $L/d > 1$ is sufficiently distant from the orifice to be drawn in it during suction stroke. For $L/d < 1$, a vortex ring formed during blowing stroke is either reingested into the orifice, or the adverse pressure gradient induced by the following suction stroke causes a rapid vortex breakdown immediately downstream from the orifice. This results in an increases in pressure required to draw new fluid into the jet cavity. When the vortex ring advection is limited by the top plate, however, L/d has to be considerably higher than $L/d = 1$ to prevent the vortex from restricting the entrainment of new fluid into the jet cavity. Following the impingement, the vortex ring stetches radially outward and depending on H/D may induce flow recirculation downstream from the orifice. A further decrease in H/D can result in an ingestion of the vortex into the jet cavity regardless of the magnitude of L/d, thereby completely blocking the entrainment. Therefore the effect of larger L/d in terms of increased entrainment and prevention of recirculation is more pronounced for higher H/D. To clarify this point the mean flow fields for $H/D = 0.04$ and $H/D = 0.08$ and two synthetic jets having the same Reynolds numbers ($R_e = 3300$) but different dimensionless stroke lengths: $L/d = 5$ (jet at 240 Hz) and $L/d = 15$ (jet at 80 Hz) are shown in Figures 7(a-d). For $H/D = 0.04$, both vortices are limiting the entraiment of fresh ambient fluid, Figures 7(a-b). When the spacing between the plates is doubled, the obstruction of ambient air entrainment is significantly reduced, Figures 7(c-d). Upon impingement, vortex stretches radially further for the lower (Figure 7d) than for the higher frequency case (Figure 7c), with vorticity levels in the boundary layer appreciably higher indicating a stronger flow along the wall for the lower frequency jet.

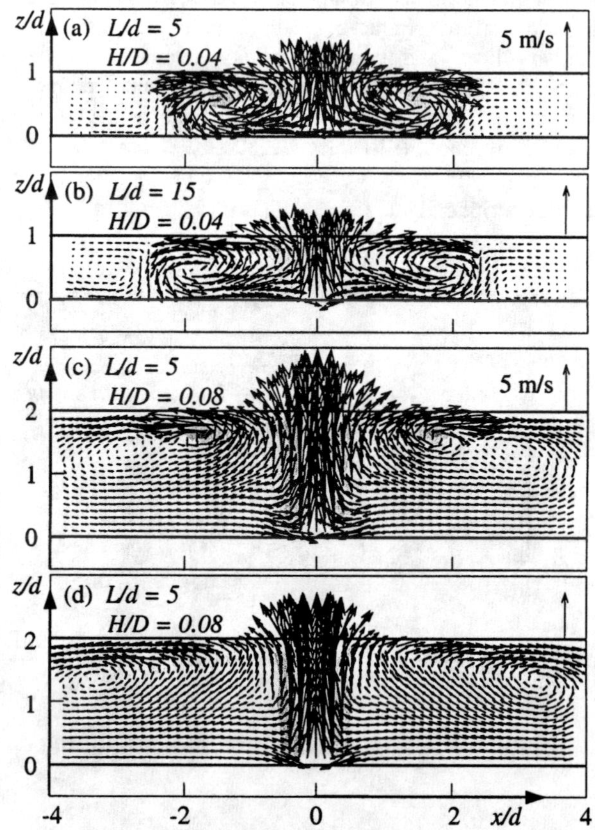

Figure 7. Comparison of time-averaged velocity distributions

IV. HEAT TRANSFER

Synthetic jet impingement induces a periodic thinning of the thermal boundary layer on the impingement surface, which decreases the thermal resistance and may improve the stagnation heat transfer. Due to the pulsatile nature of the flow, synthetic jets introduce a stronger entrainment than conventional-steady jets of the same Reynolds number and they induce more vigorous mixing between the wall boundary layers and the rest of the flow. However, it is important to recognize that although synthetic jets yield better heat convection from a solid surface to the surrounding air, they need to be formed with relatively large stroke lengths to prevent the formation of recirculating flow domains.

To further improve the heat transfer from the surface, four additional round (3.18 mm) synthetic jets are placed at a distance of 24 mm from the center jet. The spacing between the jets is sufficient to prevent their interactions and the breakdown of discrete vortex rings before their impingement on the surface. All jets operate at 80 Hz and have $L/d \approx 15$ (relative to their own orifices). The influence of peripheral jets on heat transfer is two-fold. First, the peripheral jets contribute to an overall increase in entrainment of ambient air. Second, the impingement of peripheral jets induces localized thinning of the thermal boundary layer and therefore further decreases the thermal resistance. Figure 8 shows the steady state temperature distributions along the impingement surface for the case of (i) natural convection, (ii) center jet (80 Hz, $L/d = 15$), and (iii) an array of synthetic jets driven by the same plenum, center jet and four peripheral jets (80 Hz, $L/d \approx 15$).

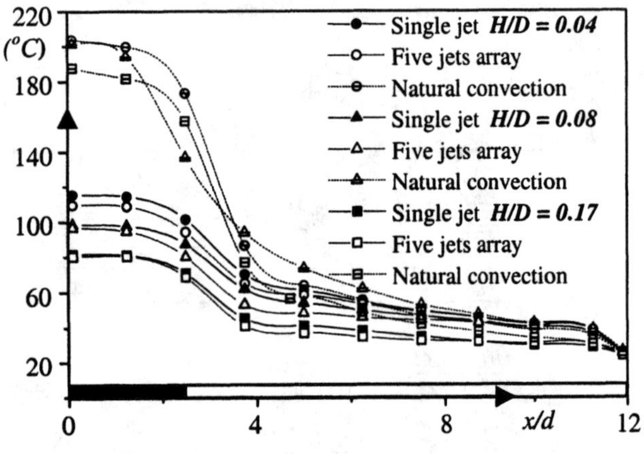

Figure 8. Surface temperature distributions

The computed global heat transfer coefficient is shown in Table 1, taking into account the conduction losses through the top surface, thermocouples and power leads, and radiation losses. The heat transfer coefficient is based on the highest temeperature difference, between the heater and the ambient, and the area of the heater. It decreases with decreasing H/D, but still shows that even in confined configurations where the airflow towards jet is restricted the jet can still effect significant cooling (although the heat transfer coefficent for $H/D = 0.04$ is approximately 60% of the coefficient for $H/D = 0.17$). The addition of peripheral jets, incrases the heat transfer coefficient but the magnitude of that improvement is highly dependent upon the flow dynamics in the gap. It is noteworthy that for the measurements reported here, the peripheral jets are not impinging on the heater but on the top plate at a distance of about d from the radial edge of the heater. It is anticipated that when the cirular plates are closely spaced, the addition of periphereal jets will maintain entrainment of ambient air when the flow towards the center jet is severely restricted.

Table 1. Global heat transfer coefficient

H/d	*1*	2	*4*
Center jet, *h* (W/m^2K)	63	79	105
Five jet array, *h* (W/m^2K)	67	80	111
Natural convection, *h*(W/m^2K)	4	5	19

V. RADIAL COUNTERCURRENT HEAT SINK

The findings from the study of the baseline (parallel plate) geometry are extended to the heat transfer investigation of an active radial countercurrent heat sink that utilizes this baseline configuration and incorporates triangular fins between the parallel plates. The heat sink diameter is 76 mm and its height is 12.7 mm, Figure 9. The triangular fins start at about 11 mm from the heat sink centerline and form 32 rectangular channels (6.35 mm in height) that spread toward the radial edge of the heat sink. Heat sink is made from aluminum and integrated with a central synthetic jet and an azimuthal array of 32 round jets. All jets are driven by the same plenum. The orifice plate is also made from aluminum and insulated on the back to minimize conduction losses.

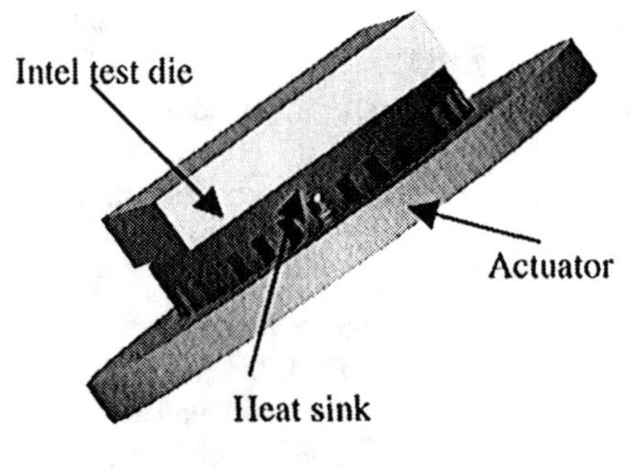

Figure 9. Sketch of the heat sink

Although the addition of triangular fins substantially increases the heat transfer surface it also increases the resistance for ambient air entrainment towards the center jet. Therefore a round peripheral jet is placed within each channel to augment the entrainment of ambient air and to contribute to the thinning of the thermal boundary layer within the channel. The presence of triangular fins leads to the azimuthal breakdown of the center jet vortex when it reaches these fins. The induced flow within each channel results in radially outward tilting of the peripheral jet and thereby facilitates the ejection of the heated fluid to ambient. Even though the heat transfer would be significantly improved if the channel height were doubled, the motivation for this design is to determine the performance of a low profile heat sink for which the radial countercurrent flow through the channels is restricted. To further restrain the flow, the orifice plate is kept at 152.4 mm in diameter, as in previous experiments.

An Intel™ flip-chip packaged thermal test die is used to assess the performance of the heat sink. The die is bonded to the heat sink and insulated on the back to maximize conduction towards the heat sink. The jets are operated at 80 Hz with a volume flow rate of about 0.17 cfm through each heat sink channel that is measured using a pitot static probe placed at the radial edge of the fins and centered in the channel. At a case to ambient temperature difference of $\Delta T_{ca} = 50$ °C, the case to ambient thermal resistance is $\theta_{ca} = 1$ °C/W. The active-heat-sink synthetic-jet module dissipates 50 W, for a nominal case temperature of $T_c = 70$ °C. Under natural convection and the same case to ambient temperature difference this system dissipates about 11 W, with thermal resistance of $\theta_{ca} = 4.5$ °C/W. This yields over four-fold improvement over natural convection cooling. Temperature distributions through the heat sink channel are given in Figure 10. The channel exit air temperature is above 40 °C suggesting that the smaller overall diameter would improve the efficiency of the heat sink. In fact shorter channels would provide even better ambient air entrainment and shorter air residence time within the heat sink.

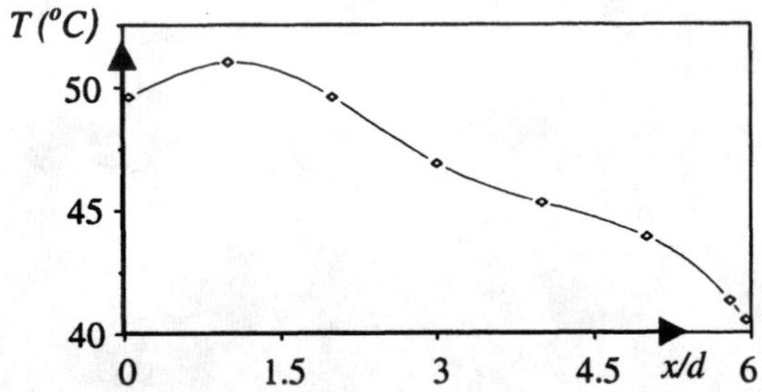

Figure 10. Air temperature distributions through the heat sink

It is noteworthy that ideally, the present heat sink could remove about 128 W, based on the heat capacity of the total airflow through the heat sink (about 5.5 cfm) and the case to ambient temperature difference ($\Delta T_{ca} = 50$ °C). With 50 W heat dissipation, this makes this module about 40% efficient in removing the ideal heat capacity. As noted by Mahalingam and Glezer (2001), an off-the-shelf active fan-heat-sink system achieves a power dissipation of 20% of its ideal capacity.

VI. CONCLUSIONS

Space and volume limitations in electronics cooling may be mitigated by air-cooling using synthetic jet impingement. This cooling approach is investigated experimentally in a confined flow geometry for which the airflow towards jet may be severely restricted. Present work focuses on the flow between two parallel plates that is induced by a centrally mounted synthetic jet. It is shown that even in the narrow gap between large parallel plates synthetic jets can effect substantial cooling by entraining cool ambient air and rejecting heated air to ambient. The primary limiting parameters for the efficient operation of synthetic jets in confined geometry are their dimensionless stroke lengths and the volume flow rate that they induce. This work has shown that in order to prevent hot air recirculation and decrease the losses for ambient air entrainment, synthetic jets having dimensionless stroke lengths of at least $L/d = 15$ may be needed in order to achieve meaningful heat dissipation.

A low profile, radial countercurrent heat sink having a diameter of 76 mm and height of 12.7 mm was tested. The heat sink dissipates 50 W at a case to ambient temperature difference of $\Delta T_{ca} = 50$ °C yielding over four-fold improvement in heat transfer compared to natural convection cooling. Furthermore, this improvement is achieved by using 5.5 cfm through the heat sink. The present data shows that even though the flow towards jet is restricted (e.g., by the gap height) the heat sink can still dispense heat to ambient more efficiently than comparable off-the-shelf heat-sink-fan air cooling modules.

ACKNOWLEDGMENTS

This work is supported by the DARPA HERETIC Program.

REFERENCES

Azevedo, L. F. A., Webb, B. W., and Queiroz, M., "Pulsed Air Jet Impingement Heat Transfer", Experimental Thermal and Fluid Science, Vol. 8, pp. 206-123, 1994.

Jambunathan, K., Lai, E., Moss, M. A., Button, B. L., " A Review of Heat Transfer Data for Single Circular Jet Impingement", International Journal of Heat and Fluid Flow, Vol. 13, pp. 106-115, 1992.

Mahalingam, R., and Glezer, A., "An Actively Cooled Heat Sink Integrated with Synthetic Jets, NHTC paper 2001-20025, 2001.

Russell, G B., "Local and System-Level Thermal Management of a Single Level Integrated Module (SLIM) using Synthetic Jets", M.S. Thesis, Georgia Institute of Technology, Atlanta, GA, 1999.

Smith, B. L., and Glezer, A., "The Formation and Evolution of Synthetic Jets", Physics of Fluids, Vol. 10, 1998, pp. 2281-2297.

Smith, B. L., Trautman, M. A., and Glezer, A., "Controlled Interactions of Adjacent Synthetic Jets", AIAA paper 99-0669, 1999.

Thompson, M. R., Denny, D. L., Black, W. Z., Hartley, J. G., and Glezer, A., "Cooling of Microelectronic Devices using Synthetic Jet Technology", Proceeding of the 11th European Microelectronics Conference, pp. 362-366, 1997.

Proceedings of
2001 ASME International Mechanical Engineering Congress and Exposition
November 11–16, 2001, New York, NY
EPP-Vol. 1

IMECE2001/EPP-24702

VIBRATION-INDUCED DROPLET ATOMIZATION (VIDA) FOR TWO-PHASE THERMAL MANAGEMENT

Bojan Vukasinovic
Graduate Research Assistant
Woodruff School of Mechanical Engineering
Georgia Institute of Technology
Atlanta, GA 30332
email:gt0706a@prism.gatech.edu

Samuel N. Heffington
Research Engineer
Woodruff School of Mechanical Engineering
Georgia Institute of Technology
Atlanta, GA 30332
email:sam.heffington@me.gatech.edu

Marc K. Smith
Associate Professor
Woodruff School of Mechanical Engineering
Georgia Institute of Technology
Atlanta, GA 30332
email:marc.smith@me.gatech.edu

Ari Glezer
Professor
Woodruff School of Mechanical Engineering
Georgia Institute of Technology
Atlanta, GA 30332
email:ari.glezer@me.gatech.edu

ABSTRACT

A novel vibration induced droplet atomization (VIDA) technology is utilized in a two-phase heat transfer cell for cooling of high-power integrated circuits. VIDA is a rapid atomization process whereby discrete liquid drops (or liquid film) are placed on the surface of a diaphragm that is vibrated near or at the coupled resonance of the system. Surface waves that develop on the liquid free surface lead to rapid bursting and the ejection of secondary droplets that are propelled away from the diaphragm. The VIDA-based heat transfer cell is similar to a heat pipe in that cooling is based on the evaporation and condensation of liquid within a closed cell. However, in contrast to a heat pipe where the liquid is delivered to the evaporator by capillary transport within a wicking material, in the VIDA cell the atomized liquid droplets are propelled towards the hot surface thus eliminating the capillary transport limit that is imposed by the wicking material. The vapor condenses near an air-side heat exchanger and the condensate is delivered to the VIDA driver. The rate of atomization and therefore the heat transfer within the cell can be actively controlled and regulated. The present paper focuses on the fundamental aspects of the VIDA atomization process of a single liquid drop. It is shown that VIDA process can be either self-intensifying or self-decaying. The global features of the VIDA, spray droplet size- and velocity distributions, and their interrelation with the external driving parameters are investigated. Finally, the performance of a current prototype of VIDA heat transfer cell is discussed.

I. INTRODUCTION

Heat transfer from high-power integrated circuits (e.g., microprocessors) to the ambient is typically limited by the conduction resistance between the package and an external surface through which the heat is to be rejected, and the convection heat transport between the surface and the ambient fluid. The present paper focuses on a heat transfer cell that can transport a relatively large heat flux across reasonably small temperature differences by exploiting an internal phase-change cycle (i.e., evaporation and condensation). While the present process is similar in concept to heat transfer through a conventional heat pipe (e.g. Dunn and Reay [1]), the present approach offers two distinct improvements. First, the performance of a heat pipe is limited by the capillary transport of the condensed liquid back to the evaporator. Second, the heat removal at the heat pipe evaporator is sequential in that the liquid evaporates in the direction of the flow and therefore can lead to longitudinal temperature gradients.

The present heat transfer cell is based on vibration induced droplet atomization (VIDA) within a self-contained sealed cell (James et al. [1]). The operation of the cell is shown schematically in Fig. 1. A thin liquid layer is continuously atomized at the cold end of the cell, and as a result of the atomization the secondary droplets are propelled towards and impinge on the hot surface where they are continuously evaporated. The vapor condenses either on condenser surfaces or on the surfaces of the atomized droplets and the condensed liquid is delivered back to the driver. Heat is rejected to the ambient using a conventional air-side heat exchanger (heat sink). Therefore the VIDA cell overcomes two limitations of conventional heat pipes in that the liquid transport to the evaporator can be easily regulated and that the entire surface of the evaporator is simultaneously and nominally uniformly covered with impinging atomized droplets.

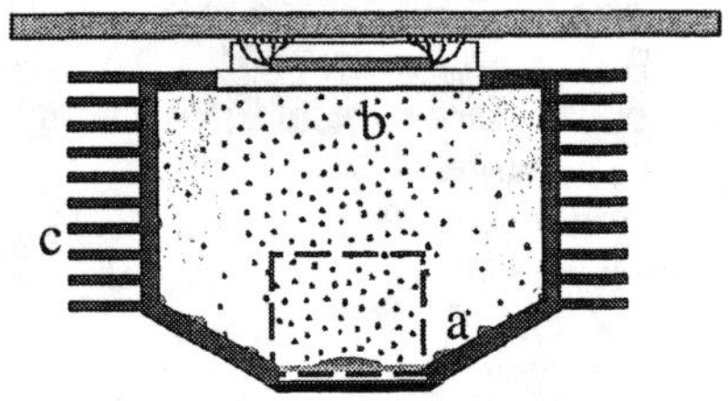

Figure 1. VIDA Heat Transfer Cell: a) VIDA atomizer, b) hot surface, and c) condensation surface.

II. THE RAPID ATOMIZATION PROCESS

While the continuous atomization within the VIDA cell typically involves a liquid film or multiple droplets, the fundamental aspects of the vibration-induced atomization are investigated using single-drop atomization. The drop is placed on a thin, circular metal diaphragm that is driven time-harmonically by a centrally-mounted piezoceramic disc. The diaphragm is typically driven at or near the coupled mechanical resonance of the transducer and the liquid droplet (or liquid layer in the VIDA cell). The relatively high oscillation frequency (on the order of 1 kHz) results in large surface accelerations (as high as 1000g) even at small displacement amplitudes.

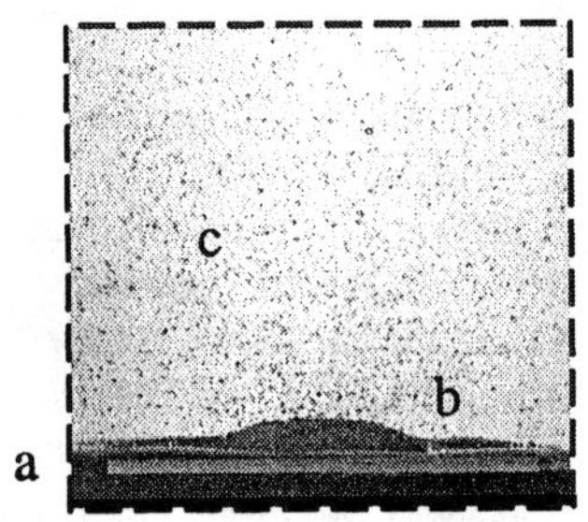

Figure 2. VIDA Atomizer: a) vibrating diaphragm, b) primary drop, and c) spray of secondary droplets.

As a result of the surface vibrations, axisymmetric standing waves are induced on the free surface of the drop. As long as the oscillation amplitude of the diaphragm is below the critical value, these surface waves are stationary. Above the critical value, the free surface becomes unstable, and subharmonic surface waves appear and grow in amplitude and complexity as the excitation amplitude increases. When the surface wave amplitude is large enough, small secondary droplets are ejected from the wave crests. As the level of forcing is further increased, this droplet ejection process intensifies until the entire primary drop abruptly breaks into a spray of small secondary droplets. One such event is presented in Fig. 2. The spray is directed away from the surface of the diaphragm, but then some of the ejected droplets fall back down onto the surface of the diaphragm due to gravity. Under certain conditions droplet ejection occurs so rapidly that the primary drop appears to atomize instantaneously. Separate experiments have shown that a similar atomization process can be achieved if a thin liquid layer is used instead of a single drop indicating that a continuous spray of secondary droplets can be created, as long as the liquid is replenished continuously.

The most important aspects of this atomization process from the standpoint of its implementation into a self-contained heat transfer cell are the characteristics of the ensuing spray of secondary droplets (the droplets velocity, size and spatial distribution) and their relationship with the operating parameters of the cell (power and frequency). This paper reports some results of the investigation of the atomization process, along with the implementation of the findings in a prototype heat transfer cell.

III. RESULTS

III.1. EXPERIMENTAL SETUP AND PROCEDURE

The design of the driver is described in greater detail in the earlier paper of James et al. [1]. The main part of the driver is a unimorph diaphragm that is constructed of a thin (32 mm diameter) metal disk bonded to a piezoceramic disk on one side. It is edge-supported in a holder and set to vibrate in its fundamental axisymmetric mode. A schematic of the experimental setup is shown in Fig. 3. The flow is illuminated by a sheet of light from an argon-ion laser. A drop is placed on the diaphragm by a micro-pipette with precision of 0.2 μl. The secondary droplet traces in a vertical plane that passes through meridian plane of the primary drop are recorded using a high-speed digital camera at (typically) 1000 fps and shutter speed of 1/20000 sec. A typical field of view for each image measures 2.2 x 2 cm (the image size is 256 x 240 pixels) and is tangential to the top of the primary drop. The dynamic system response during a droplet atomization is measured by an optical vibrometer (coupled to an A/D board) simultaneously with the high-speed imaging. The time-trace of the diaphragm velocity during an atomization event is measured with accuracy of 1.25 mm/s over the measurement range. The signal generator used for the diaphragm forcing, the vibrometer, the high-speed camera and the data acquisition board are controlled from a single PC-based LabView code. The system is forced by an stepped-amplitude sinusoidal waveform, and the dynamic response exhibits overshooting and oscillatory, exponential decay toward the final level which are characteristic of underdamped second order systems. A 0.1 ml water drop atomized nominally at 1 kHz is used in measureements of the spray velocity field. For each set of the driving parameters, 100 experimental runs are carried out.

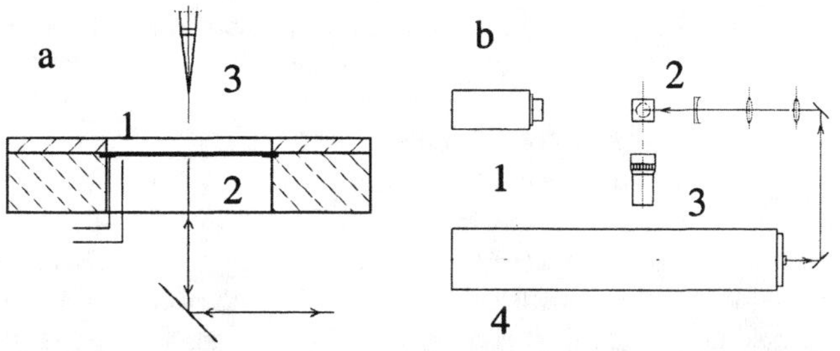

Figure 3. Experimental setup. a) actuator assembly: 1) metal diaphragm, 2) piezoceramic disk, and 3) micropipette. b) laser diagnostics: 1) vibrometer, 2) actuator, 3) high-speed camera, and 4) argon-ion laser.

III.2. VIDA GLOBAL DYNAMICS

The global characteristics of VIDA process are determined in terms of the system dynamic response to the external forcing. A typical frequency response of the dry diaphragm (i.e., in the absence of the drop) in terms of acceleration amplitude is shown in Fig. 4, for various values of the applied voltage. The behavior in the range of very weak to moderate forcing (i.e., 1g to 150g) is typical of a weakly nonlinear structural system that is forced near its resonant frequency. As the forcing amplitude increases, the nonlinearity in the system becomes more and more pronounced. For the forcing at a high level, the acceleration near resonance appears as a discontinuity in the frequency response. This is clearly seen for the frequency response curve for the highest applied forcing in Fig. 4a and is probably due to hysteresis in the response (although this was not verified experimentally). For most of this experimental investigation, experiments are carried out in the response range that excludes discontinuities. Therefore, the characteristics of the atomization process are not the result of a shock-induced acceleration due to the jump in the response curve.

The influence of the presence of a small drop of water on the diaphragm, without any ejection is shown in Fig. 4b. As expected, the additional mass of the water drop lowers the resonance frequency of the system. The peak acceleration amplitude at resonance decreases slightly with droplet mass. This behavior is typical of the structural damping that occurs as a result of the flexing of the diaphragm. A family of frequency response curves is obtained by forcing the system at a very weak level in order to maintain a stable drop surface. The nonlinear behavior of the clamped diaphragm at these low-level forcing is not pronounced.

Knowledge of the shift in the frequency response of the coupled drop and diaphragm with the mass of the drop can be used to explain the global bursting event. The most interesting observation is that the primary drop can burst while the forcing frequency and voltage are fixed.

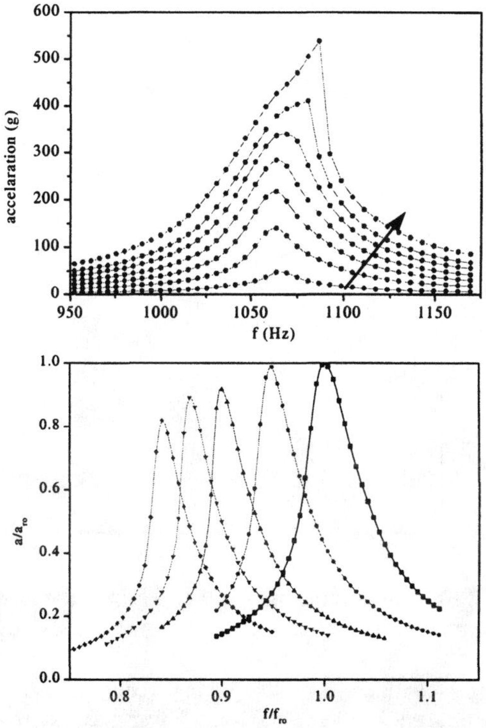

Figure 4. a) Frequency response of a dry diaphragm with applied voltage and b) frequency-shift response with the primary drop volume: ■- 0, ●- 0.1, ▲- 0.2, ▼- 0.3, ◆- 0.4 ml.

However, this bursting occurs when two conditions are met. First, the forcing parameters must be such that the response amplitude is large enough to trigger initial droplet ejection. Second, the forcing frequency must be larger than the initial resonance frequency of the coupled system. The initial weak ejection of the secondary droplets develops over time and as the mass of the primary drop decreases the ejection rate increases and ultimately results in rapid drop atomization (or bursting). This transient event may be thought of as a continuous sequence of quasi-steady resonance states as demonstrated by the shifting frequency curves in Fig. 5.

In both sketches in Fig. 5, the curves drawn with the lightest shade of gray correspond to the initial primary drop (i.e., before any ejection occurs). As noted above, when the ejection of the secondary drops begins, the mass of the primary drop is reduced and the corresponding response curve for the lower-mass system (Fig. 4a) is shifted toward a higher frequency. Four response curves from such a family are presented in each of the diagrams in Fig. 5.

Two distinct cases that depend on the ratio of the forcing frequency f_d to the initial resonance frequency f_r should be noted. When f_d is higher than f_r, (Fig. 5a) and the forcing amplitude is high enough so that secondary droplet ejection is triggered, the forcing amplitude increases with decreasing mass of the primary droplet (as indicated by the arrow), the ejection intensifies and the entire primary drop may eventually dry out.

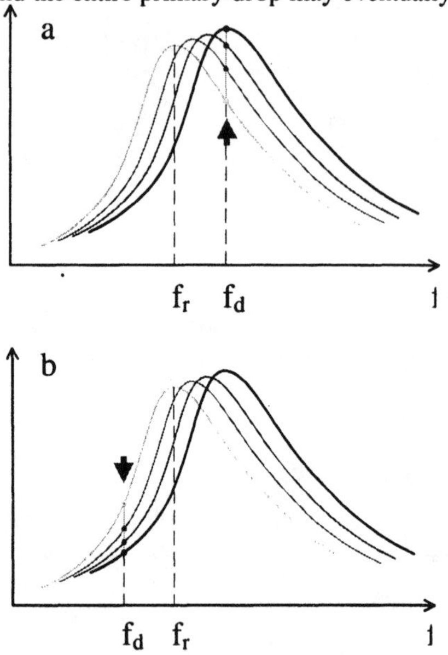

Figure 5. Ejection dynamics if the driving frequency f_d is higher (a) / lower (b) than resonant frequency f_r.

This process continues until the forcing frequency matches the resonant frequency of the transient lower-mass system (the black curve in Fig. 5a). If at that point the primary drop is not completely drained, further ejection leads to a decrease in the response amplitude, the ejection decays, and at the end of the process some fraction of the primary drop is not atomized.

However, if the forcing frequency is lower than resonant frequency of the initial system, the primary drop is not atomized (Fig. 5b). If the initial forcing amplitude is high enough to induce ejection of secondary droplets, as the mass of the primary droplet is reduced the response amplitude decreases as well and therefore, the ejection rate weakens and eventually the ejection is completely suppressed.

III.3. SPRAY SPATIAL DISTRIBUTION

One of the important characteristics of the spray of secondary droplets is the spreading of the spray and the droplet density. It is anticipated that at least initially the distribution is nominally axisymmetric (initially both the drop and the diaphragm forcing are axisymmetric). Figure 6 shows the distribution of the centroids of the spray droplets for the three instances during the atomization of the primary drop (start-up t_1, developing spray t_2, and fully-developed spray t_3). The distribution at each instant is comprised of an ensemble of 100 realizations each obtained at the same delay relative to the onset of the atomization. A single snap shot of the spray illuminated by a point light source at each of the three instances is shown in the figure. The spatial distribution of the secondary droplets in the radial and vertical direction is shown along the lower row of images. During these stages of the atomization, the spray is reasonably axisymmetric within the measurement meridian plane. As the atomization process continues, the radial distribution which is initially bipolar becomes uniform with a peak at the centerline of the primary drop. It is noted that the initial bipolar distribution is in agreement with the high-speed video imaging that shows that the atomization starts along the contact line of the primary droplet. These images also show that shortly after that beginning of the atomization, the primary

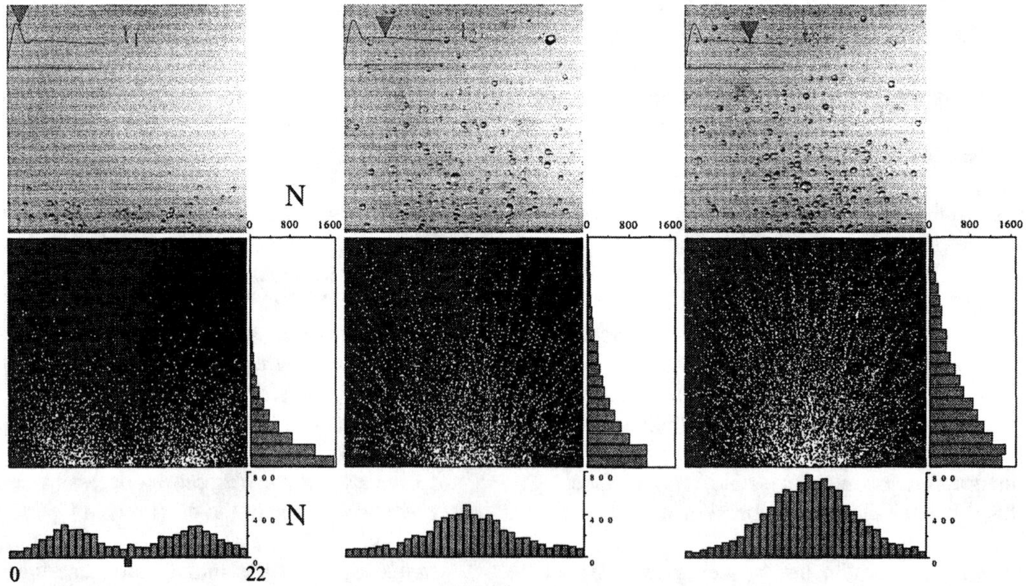

Figure 6. Drop atomization: upper row - single-image visualization, lower row - overlapped centroids of the spray droplets.

drop is atomized over its entire surface (e.g., at times t_2,) until the primary drop is drained out. A primary drop having a volume of 0.1-ml is typically atomized within 0.3 sec.

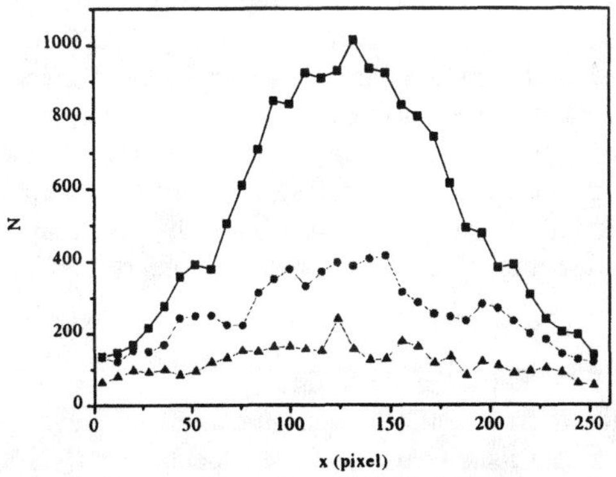

Figure 7. Secondary droplet distribution at: ■ - 10, ● - 15, and ▲- 20 mm above the actuator.

Finally, secondary-droplet distributions at three levels above the transducer during the atomization are shown in Fig. 7. As expected, the spatial non-uniformity of the spray decreases with height. It can be concluded that a (co-axial) circular target up to some 12 mm in diameter can be uniformly coated if placed approximately 15 mm above the transducer.

III.4. DROPLET SIZE DISTRIBUTION

The high-speed images of the spray enable measurements of the projected cross sectional area of the droplets and an assessment of their mean diameter. Depending on the driving frequency, the image resolution is adaptively increased along with the frequency. Note that the thickness of the illuminating light sheet is comparable with diameter of the droplets and requires an adjustment of the threshold-intensity for processing of the raw images.

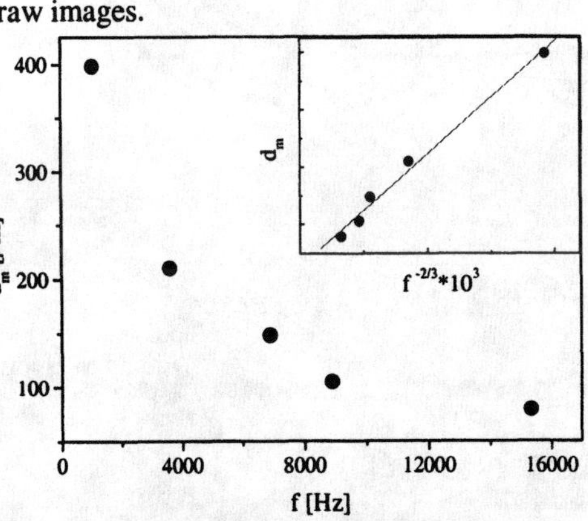

Figure 8. Mean diameter of the secondary droplets with the driving frequency.

In the present set of measurements, water drops ranging from 10 to 150 μl are atomized by VIDA actuators operating nominally at 1, 3, 7, 9, and 15 kHz. The corresponding range of image resolution varied between 25 and 9 μm/pixel. The distributions of the mean diameters of the secondary-droplets are shown in Fig. 8 (the standard deviation is 40%). These data show that for a given fluid of fixed properties and within the resolution of the present measurements, the mean diameter of the secondary droplets is invariant with the size (volume) of the primary primary drop during the atomization process. The only parameter that strongly affects the size of the secondary droplets is apparently the driving frequency that determines the dominant wavelength on the free surface of the primary drop. This wavelength varies like $f^{2/3}$ and therefore it is expected that the spray mean diameter varies like $f^{-2/3}$. Figure 8 shows that the mean droplet size distribution is in good agreement with this frequency dependence.

III.5. SPRAY VELOCITY

The velocities of the spray droplets are extracted from the series of high-speed images using a three-frame particle-tracking algorithm (Vukasinovic et al. [3]). The time-averaged velocity fields corresponding to the three characteristic instances during the spray evolution are shown in Fig. 9.

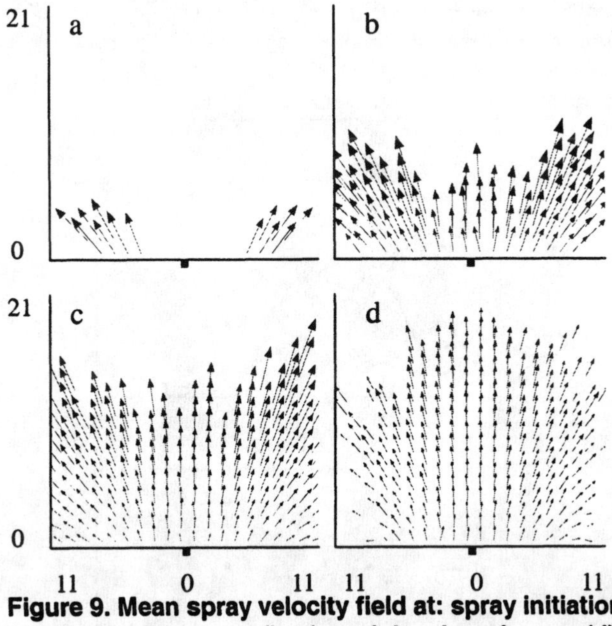

Figure 9. Mean spray velocity field at: spray initiation (a), developing spray (b, c), and developed spray (d).

The velocity is averaged within a 12 x 12 pixel spot, and it is based on no less than 15 vectors. During the start-up of the atomization (Fig. 9a-b), the spray velocity appears to have of higher magnitude than in later stages of the atomization.

Following the brief transient that is associated with the startup of the atomization, it appears that the spray droplet velocity is comparable with the velocity of the vibrating diaphragm (Fig. 9c-d). This conclusion is in agreement with the findings of Yule and Al-Suleimani [4] who measured the velocity of the droplets ejected from a vibrated column of liquid. In the present measurements, the initial velocities of the spray droplets can be as high as 3 m/s, and are typically in the range of 0.5-1 m/s during the rest of the process.

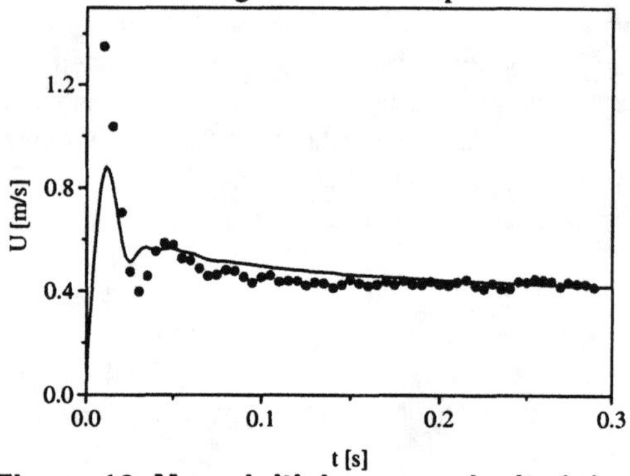

Figure 10. Mean initial spray velocity (●) with diaphragm velocity (—) over time.

These results suggest that the spray velocity depends not only on the forcing level but also on its rate of change. The variation of the initial velocity within a 12-pixel strip above the primary drop with the corresponding diaphragm velocity (Fig. 10) shows that the transient (in this case stepped) start-up of the forcing enables higher initial velocities of the spray. This suggests that the spray velocity can be controlled by proper modulation of the driving signal.

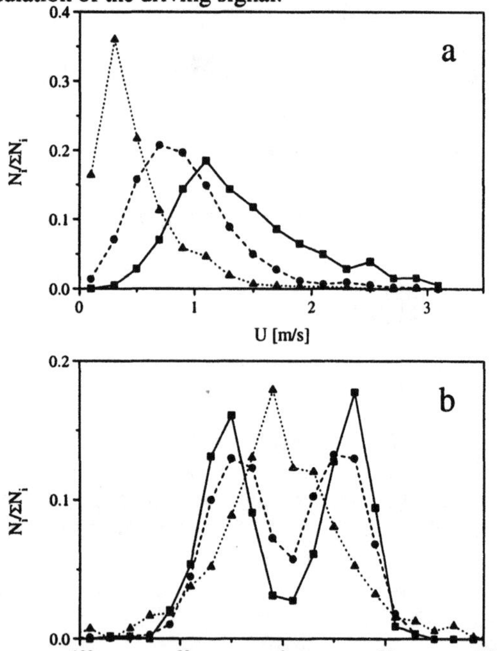

Figure 11. Initial mean spray velocity distribution (a) and distribution of the velocity vectors over angles (b) for the spray initiation (—), spray development (---) and developed spray (···).

The distributions of the velocity magnitude and direction (i.e., angle relative to the centerline) during the atomization process are plotted in the Fig. 11 for the three characteristic instances during the atomization (Figure 6). Along with the shift toward a lower velocity as the atomization process develops, it can be seen that the deviation from the mean value becomes less prominent. The initial velocity distribution over angles matches the conclusions drawn from the discussion of spatial distribution in Section III.3 above. As noted in Section III.3, because the atomization begins at the contact line of the primary drop, the distribution during the startup is bipolar, and as the process evolves, it blends into the unipolar, fairly symmetric distribution around the centerline of the primary drop.

IV. VIDA HEAT TRANSFER CELL

Several different VIDA cells have been constructed and tested to evaluate their heat transfer characteristics (Heffington et al. [5]). A schematic of one of the recent VIDA heat transfer cell designs is shown in Fig. 12. The vapor space in the top view has a cross-shaped section, and the driver is mounted in the bottom of the cell. The calibrated heater is mounted in the top of the cell and circular, equally-spaced fins are located around the external portions of the cell. A conventional microelectronics fan is positioned below the fins, and it is used to circulate cool air over the fins surfaces.

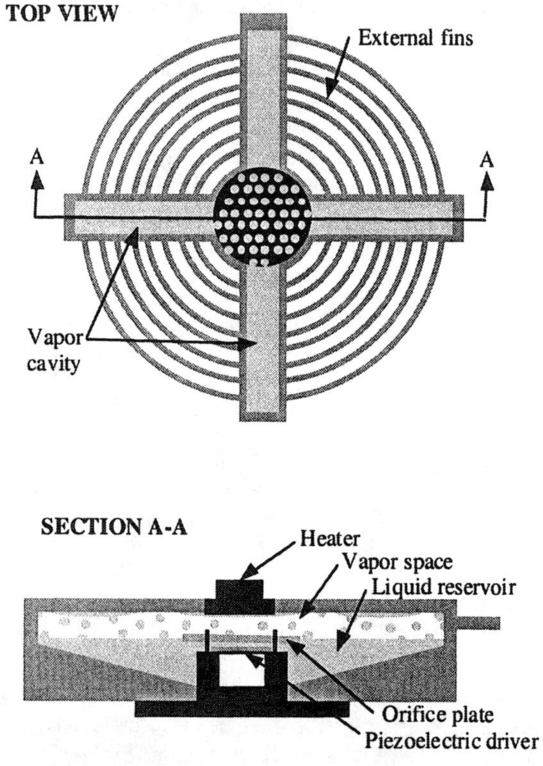

Figure 12. Schematic of the VIDA heat-transfer cell.

The vapor space is designed to provide a large condensation area so that the base of the fins is maintained at a temperature close to the saturation temperature of the water. The base of the vapor cavity is slanted so that the condensed liquid is routed back to the driver so that there is a thin liquid layer of uniform thickness over the driver. The atomization is achieved using a 15 mm diameter piezoelectric driver mounted below a perforated orifice plate having uniformly spaced holes. Besides atomization, the motion of the driver is also used to pump liquid from the thin layer towards the actuator's center through a small gap between the driver and orifice plate. The pressure of both sides of the driver is equalized (when operated at reduced pressure) using a connecting tube to the vapor space. The VIDA process is monitored and tuned (in terms of operating frequency and liquid volume) by replacing the calibrated heater with a clear window. For the present cell (Fig. 12), a strong VIDA process is achieved when the cell was charged 5-6 mL of water and driven at 2.3 to 2.6 kHz.

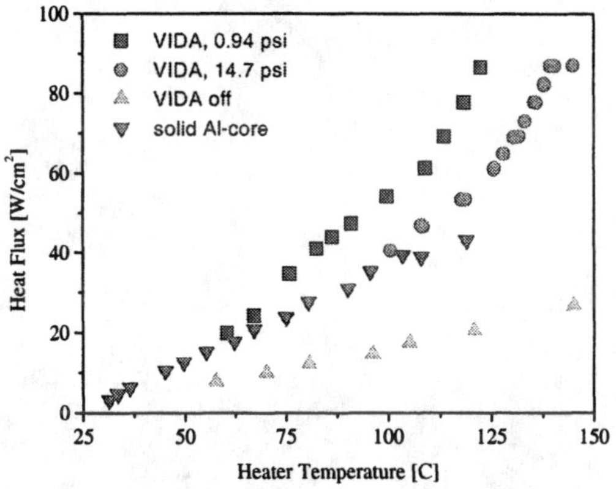

Figure 13. Heat transfer performance of the VIDA cell

The heat transfer rate removed at the heater surface as a function of heater surface temperature is shown in Figure 13 (when the cell is operating at 2.35 kHz). It is noteworthy that the heat transfer results for this cell are similar to what is achieved in conventional pool boiling curve (Nukiyama [6]). The heat flux increases with temperature up to a critical level at which a further increase in temperature leads to a reduction in the heat flux owing to the formation of an insulting vapor layer. In the present measurements, the critical heat transfer rate was 248 W at a heater surface temperature of 140°C. (Several tests at this power level verified the repeatability of these results.) However, it is evident that the required surface temperatures are in excess of the range of interest for most microelectronic applications.

To achieve an operating temperature that is more compatible with microelectronic cooling applications, the VIDA heat transfer cell is operated at reduced pressures. By lowering the operating pressure within the cell, the saturation temperature of the working liquid is decreased thereby allowing the reduction of the heater temperature to acceptable levels. The pressure within the cell is initially set to 0.94 psia, for which the saturation temperature of water is 37.3°C (the operating frequency of the transducer is 2.9 kHz). The pressure within the cell increases slightly with heater temperature, but after the completion of the test, the pressure is to just less than 1 psia, indicating that the cell is pressure tight. Since the pressure is reduced, the critical heat transfer (250 W) is not significantly different from the what is achieved at atmospheric pressure, but the surface temperature at critical heat flux is of 122°C, or 18°C lower than under atmospheric conditions. Although a higher level of excess temperature is required at reduced pressure, the overall thermal resistance of the cell is reduced.

The results of the atmospheric and reduced pressure tests are compared to tests in which the VIDA process is inactive (i.e., in the absence of evaporation) in order to demonstrate its effectiveness. The first heat transfer test in the absence of VIDA used a solid aluminum core that replaced the vapor space within the cell. In the second test, the hollow vapor space was evacuated but the cell was not charged with water. It is shown that the droplet atomization process improves the thermal performance of the cell compared to the heat flux for a solid aluminum cell, and this improvement is more evident at elevated temperatures.

V. CONCLUSIONS

A new VIDA process is described for active dispersion of liquid droplets within a two-phase heat transfer cell. The present investigation focuses on global features of the VIDA process with emphasis on the process parameters that are relevant for its practical implementation. Rapid liquid atomization is enabled by a dynamic interaction between a vibrating driver surface and the adjacent liquid when the driving frequency is larger than resonance frequency of coupled driver fluid system. During the atomization process, the ensuing spray of secondary droplets is self-propelled and reasonably symmetric about the driver centerline. In the present experiments, the standard deviation of the spray droplets is nominally 40% of the mean diameter. The mean diameter depends strongly on the driving frequency, and varies like $f^{-2/3}$. The initial velocity of the spray depends on both the level of the forcing and its rate of change.

Initial heat transfer results utilizing water and a simple cell design have shown that a cell based on the VIDA process can provide a heat flux of nearly 100 W/cm^2. Reducing the pressure inside the cell increases the heat flux at a given heater surface temperature. The VIDA cell outperforms an identical cell consisting of a solid aluminum core. These preliminary tests have indicated that a VIDA-based heat transfer cell has a potential to provide inexpensive cooling of small-scale electronic devices and the cooling capabilities are within the range required for most modern package designs.

ACKNOWLEDGMENTS

This work is supported in part by the NASA Microgravity Research Division and the DARPA HERETIC program.

REFERENCES

[1] James, A., Vukasinovic, B., Smith, M.K., and Glezer, A., 2001, "Vibration-induced droplet atomization and drop bursting", submitted to Journal of Fluid Mechanics.

[2] Dunn, P. D. and Reay, D.A., 1994, *Heat Pipes*, Elsevier Science Ltd.

[3] Vukasinovic, B., Smith, M.K., and Glezer, A., 1999, "An Experimental Investigation of Vibration Induced Droplet Atomization", Bulletin, APS/DFD99, **44**, pg. 26.

[4] Yule, A.J., and Al-Suleimani, Y., 2000, "On droplet formation from capillary waves on a vibrating surface", Proceedings of the Royal Society of London A, **456**, pp. 1069-1085.

[5] Heffington, S.N., Black, W.Z., and Glezer, A., 2000, "Vibration-induced droplet cooling of microelectronic components", Proceedings, ITHERM 2000, **2**, pp. 328-332.

[6] Nukiyama, S. 1934 "The maximum and minimum values of heat transfer from metal to boiling water under atmospheric pressure," J. Japan Soc. Mech. Eng., **37**, 367, (Translation: 1966 Int. J. Heat Mass Transfer, **9**, 1419).

Proceedings of
2001 ASME International Mechanical Engineering Congress and Exposition
November 11-16, 2001, New York, NY
EPP-Vol. 1

IMECE2001/EPP-24703

Bumpless Build-Up Layer Packaging

Steven N. Towle*, Henning Braunisch, Chuan Hu, Richard D. Emery, and Gilroy J. Vandentop
Intel Corporation
Components Research
5000 West Chandler Boulevard
Chandler, Arizona 85226-3699, USA

Abstract—**Bumpless Build-Up Layer (BBUL) is a novel package developed to meet future packaging technology requirements. The BBUL package provides the advantages of small electrical loop inductance and reduced thermomechanical stresses on low dielectric constant (low-*k*) die materials. Furthermore, it allows for high lead count, ready integration of multiple electronic and optical components [such as logic, memory, radio frequency, microelectromechanical systems (MEMS), among others], and inherent scalability. In the present paper we investigate and discuss some of the process, routing, electrical, thermal, and mechanical attributes of BBUL.**

1. Introduction

As integrated circuit technology advances, packaging high performance microprocessors becomes increasingly challenging. Future generations of Intel microprocessors are expected to have larger numbers of signal leads, stricter control of impedance and crosstalk on these lines, and greater demands for power delivery and heat extraction. For microelectronic packaging, two of the major challenges are electrical performance and thermal management. From an electrical standpoint, the goal of the packaging world is to reduce the signal integrity and operating frequency of the semiconductor device as little as possible. This task is complicated by the often large inductances introduced to the total device-package-motherboard circuit by the package. On the other hand, the package is almost completely responsible for providing a means for extracting thermal energy dissipated at the active semiconductor area.

Aside from the electrical and thermal performance considerations, other technical challenges remain. Continued reduction in end product size requires a package with a small form factor that allows for dense placement of input and output electrical connections. Future microprocessors may also require integration of multiple chips or other electrical and optical components on the same package with minimum component spacing, maximum number of interconnects between components, and tight restrictions on signal latency, impedance matching, and noise. Furthermore, the cost budget available for packaging and assembly is flat or declining, following the trend of microprocessor average sales prices. Meeting these demands stretches or exceeds the capabilities of current packaging technologies, such as Intel's flip-chip pin grid array (FCPGA) package [1].

An added challenge for future generations of logic technology is presented by the anticipated integration of low dielectric constant (low-k) materials on-chip. Low-k dielectric materials, especially those with dielectric constants at or approaching a value of 2, have very poor mechanical robustness and will be able to withstand little or no tensile or shear force. Mismatch of coefficients of thermal expansion (CTE) among the various component materials of a package makes satisfying this requirement difficult.

Many of these issues could be addressed with a chip-scale package (CSP) in which the package redistribution layers are built up on top of the die [2]–[5]. However, packaging advanced logic chips makes additional demands that require the area of the package to be considerably larger than the area of the chip. Area is needed for an adequate number of pins for the signals and power while meeting motherboard pitch limitations, and for added components such as decoupling capacitors. To meet these myriad requirements, we are developing a new package, the Bumpless Build-Up Layer (BBUL) package.

BBUL is designed to meet packaging technology requirements for Intel's 65 nm generation silicon technology and beyond. Microprocessors produced with this generation of silicon process technology are expected to reach speeds in excess of 10 GHz. BBUL provides advantages of low loop inductance for decoupling capacitors, high lead count, ready integration of multiple electronic and optical components [e.g., logic, memory, radio frequency, microelectromechanical systems (MEMS), etc.], low mechanical forces imposed on the die, and inherent scalability. BBUL does not specifically address the issue of heat extraction, but it is compatible with future advances in thermal systems that may be developed separately. In this paper we demonstrate the routing advantages provided by BBUL, as well as simulations showing significant improvement in electrical power delivery performance for BBUL vs. a standard FCPGA package. Reliability results and simulations demonstrating the mechanical

*E-mail: steven.towle@intel.com

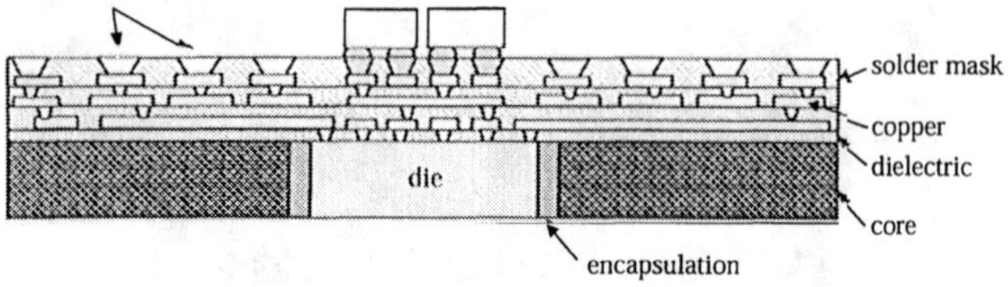

Figure 1: Schematic cross-section of a three-layer BBUL package.

advantages of BBUL for integration of on-chip low-k dielectric material are presented elsewhere.

2. Process and routing

BBUL differs from traditional assembled packages in that it consists of a die or dice embedded in a substrate, such as bismaleimide triazine (BT) laminate or a copper heat spreader, which then has one or more build-up layers formed on top. A standard microvia formation process, such as laser drilling, makes the connections between the build-up layers and the die bond pads. This is analogous to a wafer level CSP (WLCSP) with the die embedded in the panel to increase the area. The embedding of the die or dice in the panel may be done with molding or dispensed encapsulation material. As with certain other proposed CSP technologies, the build-up layers are made with a standard high-density integration (HDI) patterning technology. This build-up technology is similar to what is typically used for advanced organic packages (e.g., FCPGA) and printed circuit boards. Die bond pads may be arranged in any pattern subject to the pitch limitations of the HDI process. To simplify the laser drilling process and to relax the via to bond pad alignment constraint, a copper pad expansion may be formed over the bond pad openings, as shown in the schematic cross-section of Fig. 1. In Fig. 2 we present photographs of a BBUL-packaged test chip [Fig. 2(a)] and this same package in cross-section [Fig. 2(b)]. The process flow is illustrated in Fig. 3.

The BBUL structure offers several routing advantages compared to flip-chip interconnection. First, the current via size and alignment capabilities of HDI via formation processes allow a tighter pitch for the die-package interconnections compared to flip-chip. Continuous improvements in HDI substrate patterning capabilities are driven by the requirements of a broad range of packaging and circuit board applications; BBUL die-package pitch can be scaled down as these improvements come on line. Also, unlike many versions of flip-chip assembly, die-package interconnections can be arbitrarily placed, because no restriction is imposed by limitations of the flip-chip underfill process. This capability provides a significant advantage in the number of signals that can be routed out from the die on a single layer, as shown in Fig. 4. Since the BBUL package is one-sided, there are no plated through-holes and no underutilized wiring layers on the backside of the package. The combination of these effects can typically reduce the required layer count by 50% or more. Table 1 shows some typical routing results comparing the current state-of-the-art for flip-chip and BBUL.

Due to the layer count reduction and elimination of the package core and die-package solder connection, BBUL technology allows the overall package to be very thin. The thickness of the package will exceed the die thickness by 100 μm or less, depending on the thermal and socketing solution used. This attribute is attractive for mobile (laptop) applications and is also a main contributing factor to the improvements offered by BBUL in power delivery and mechanical stress.

Of course, BBUL is subject to the same drivers for known good die (KGD) yield and assembly and packaging yield that apply to other build-up CSPs. The KGD issue can be addressed by testing the dice before they are embedded in the package panel. Yield loss in the packaging steps result in loss of the known good dice; this die yield loss adds to the overall cost of the package. However, we believe that, with appropriate yield optimization of

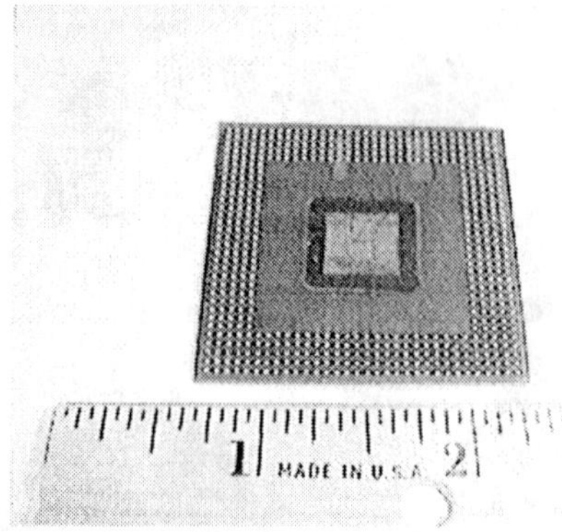

(a) Test chip in BBUL package with land grid array.

(b) Cross-sectional view of die-package interface.

Figure 2: Photographs of a BBUL package.

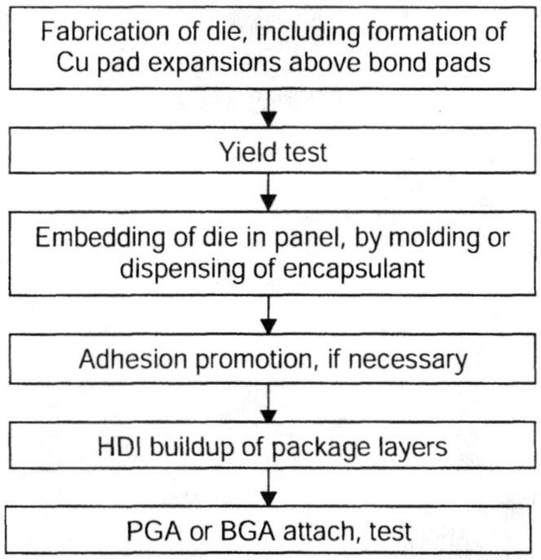

Figure 3: Process flow for formation of BBUL package.

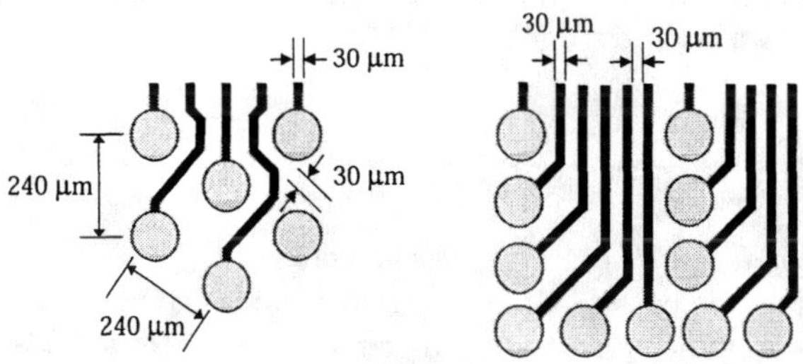

(a) Standard package (constrained to fixed pitch).

(b) BBUL package with equivalent design rules.

Figure 4: Escape routing.

the package processing, the performance advantages combined with the package layer count reduction should allow this package to be cost effective in the advanced microprocessor market segment.

3. Electrical performance

In order to demonstrate and quantify some of the electrical performance enhancements offered by the BBUL technology, we conduct transient electromagnetic (EM) simulations for the core power delivery problem. We start out with a simplistic lumped-element theory, setting the stage for the EM simulations that consider realistic package designs and providing definitions of figures of merit used in the interpretation of the simulation results.

	Flip-chip	BBUL
Minimum pitch		
Bump pitch [μm]	240	No bumps
Build-up land size [μm]	150	150
Minimum land spacing [μm]	30	30
Minimum die-package pitch [μm]	240	180
Escape routing		
Core pitch [μm]	335×335	250×250
Periphery pitch [μm]	240×415	≥180
Build-up line/space width [μm]	30/30	30/30
Build-up land size [μm]	150	150
Escape pitch [μm]	100 (4 rows)	75 (4 rows)

Table 1: Comparison of routing capabilities of flip-chip and BBUL, based on equivalent design rules (Intel 0.13 μm generation process technology for flip-chip and package).

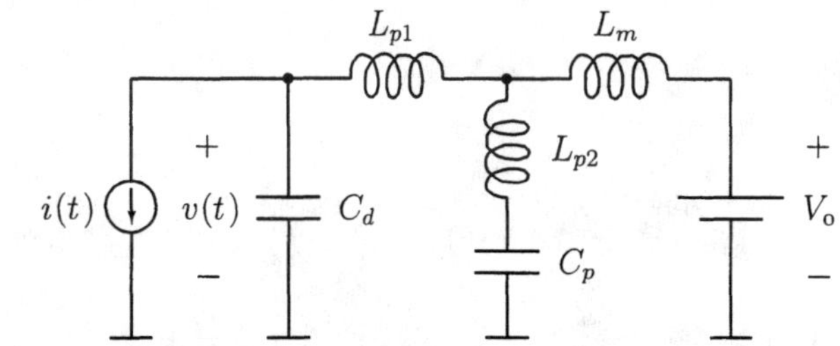

Figure 5: Lumped-element model of a power delivery structure.

Lumped-element power delivery modeling

Figure 5 shows a simplified lumped-element model of a power delivery structure involving an active die, a package, and a DC voltage supply [6]. For a quiet die, $i(t) = I_0 = \text{const.}$ with I_0 a leakage current flowing through the package inductance L_{p1} and the inductance L_m in between the package and the power supply located on the motherboard. The supply voltage V_0 appears as $v(t)$ across the on-die, high-frequency decoupling capacitance C_d that is inactive in the DC case, similar to the on-package, mid-frequency decoupling capacitance C_p with its associated parasitic inductance L_{p2}. Simultaneous switching of a large number of on-die core devices will draw a large current from the package which, in a simplistic manner, can be modeled as

$$i(t) = I_0 + \Delta I\, u(t) \tag{1}$$

where ΔI is the current step and $u(t)$ the unit step function. During the first few nanoseconds following the current step the relative large inductance L_m effectively isolates the power supply from the rest of the circuit and, letting $L_m \to \infty$, the inductances L_{p1} and L_{p2} combine as

$$L_p = L_{p1} + L_{p2} \tag{2}$$

where L_p is referred to in the following as the loop inductance of

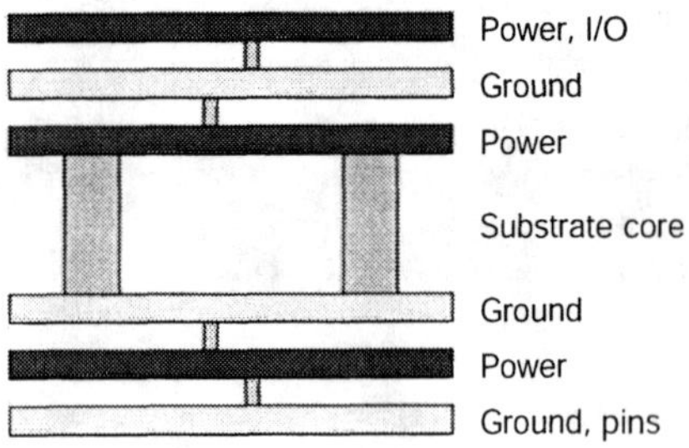

Figure 6: Stack-up of a six-layer standard package.

the package. Denoting

$$\omega_0^2 = \omega_1^2 + \omega_2^2 \tag{3}$$

where

$$\omega_1^2 = \frac{1}{L_p\, C_d}\,, \quad \omega_2^2 = \frac{1}{L_p\, C_p} \tag{4}$$

the initial-value problem is readily solved using the Laplace transform formalism and we obtain

$$v(t) = V_0 - \frac{\Delta I}{\omega_0\, C_d} \left[\left(1 - \frac{\omega_2^2}{\omega_0^2} \right) \sin \omega_0 t + \frac{\omega_2^2}{\omega_0} t \right] u(t) \tag{5}$$

Equation (5) describes the combination of oscillations of the resonator formed by decoupling capacitance and loop inductance on one hand, and the progressing discharging of the capacitors on the other. For typical microchips C_p is much larger than C_d. As $C_p \rightarrow \infty$, we find that Eq. (5) simplifies to

$$v(t) = V_0 - \Delta I \sqrt{\frac{L_p}{C_d}}\, \sin \omega_0 t\; u(t) \tag{6}$$

where

$$\omega_0 = \frac{1}{\sqrt{L_p\, C_d}} \tag{7}$$

The "drooping" of the die voltage as described by Eq. (5) and in simplified form by Eq. (6) can limit the performance of the microchip.[1] Equation (6) displays the dependence on loop inductance explicitly. BBUL targets the reduction of the inductance L_{p1} in Fig. 5 by reducing the thickness of the package. This in turn may allow usage of a smaller on-package capacitance C_p with lower associated parasitic inductance L_{p2}. Thus, the loop inductance L_p of the package can be reduced significantly. In the following section we demonstrate this by simulations based on an actual package design, quantifying the improvement in loop inductance BBUL is expected to provide.

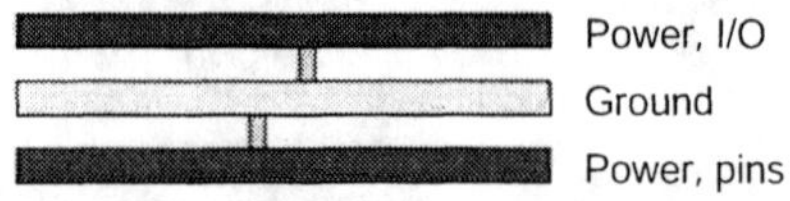

Figure 7: Stack-up of a three-layer BBUL package.

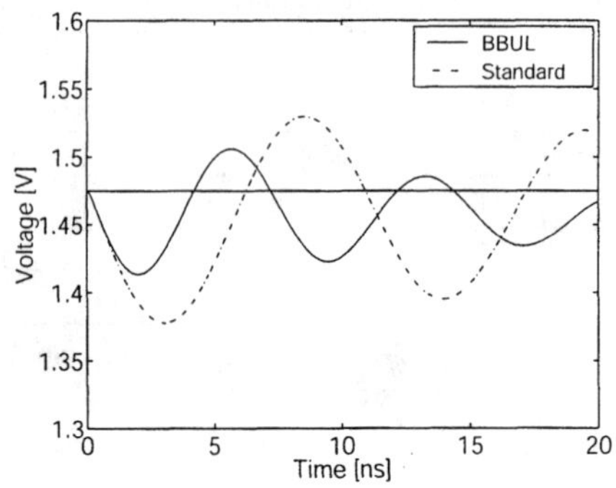

Figure 8: Transient simulation results for the mean die voltage of the two packages.

Design-driven electromagnetic simulation

We obtain power delivery time-domain results for a standard six-layer flip-chip package and compare them with a model of a similar three-layer BBUL package. The simulations are carried out using the commercial Sigrity SPEED2000^{TM} tool [8], incorporating EM wave propagation through the multiple layers of the complex package structures. The overall methodology employed allows a realistic comparison of the electrical performance of BBUL against a standard package for a major Intel product.

The power and ground nets of the standard package were extracted from the full design file and translated into the EM simulator format. Careful manual and semi-automatic editing was then applied to generate a structure that can be simulated successfully. The resulting model contains power and ground planes of various shapes and more than 4200 vias; the individual locations of these vias are fully taken into account during the simulation. The schematic stack-up structures of the two packages are shown in Fig. 6 and 7, respectively, indicating the basic functionality of the different layers. The BBUL package was generated by retaining the two top layers and the bottom layer of the standard package, reversing the polarity of the bottom layer, and reconnecting the vias as appropriate. In this BBUL package based on a realistic design we encounter about 2300 vias. A distributed 5×5 die model similar to what is shown in [7] and incorporating a power map is connected to the packages. Land-side decoupling capacitors amounting to a C_p of more than 20 μF are placed at the bottom of the package. The lumped parasitic inductance of these

[1] It is pointed out that the analysis holds only for the early portion of the response, as indicated by the unbounded behavior of Eq. (5). A correct description at later times needs to take into account losses and, more importantly, the influence of the power distribution system beyond the boundaries of the package [7].

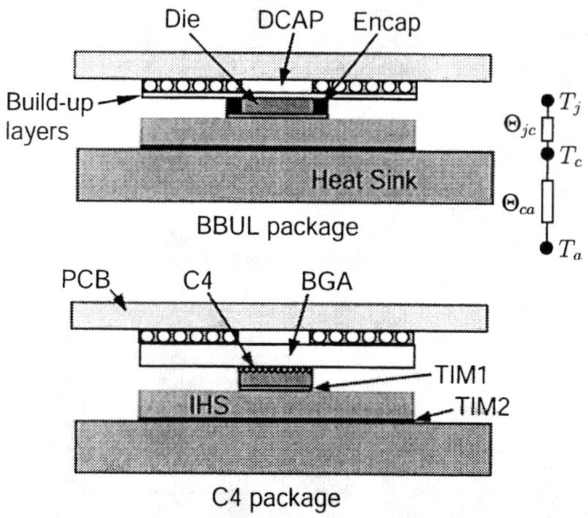

Figure 9: Thermal models of BBUL and C4 package.

	$\kappa_{th,x}$, $\kappa_{th,y}$, $\kappa_{th,z}$ [W/mK]	L_x, L_y, L_z [mm]
Sink	398, 398, 398	64, 64, 6.4
TIM2	30, 30, 30	34, 34, 0.2
IHS	398, 398, 398	34, 34, 15
TIM1	64, 64, 64	11.9, 11.9, 0.2
Die	120, 120, 120	11.9, 11.9, 0.775
Encap	0.2, 0.2, 0.2	13.9, 13.9, 0.775
Build-up layers	3, 3, 0.69	34, 34, 0.135
C4 bumps	0.5, 0.5, 1.24	11.9, 11.9, 0.1
BGA board	50, 50, 2	34, 34, 1.1
Solder balls	5, 5, 28	34, 34, 0.5
DCAP	0, 0, 0	11.9, 11.9, 0.5
PCB	3, 3, 0.3	64, 64, 1.6

Table 2: Thermal input parameters: thermal conductivities and dimensions of the different regions.

capacitors is given by L_{p2} =1.3 pH. Finally, an ideal power supply is connected directly at all the power and ground pins; there is no socket or motherboard in the simulated models. Thus, the inductance L_m in Fig. 5 is formed by the lateral plane inductance of the packages; the on-package decoupling capacitors are confined to the area directly underneath the die, whereas the pins are along the periphery of the package.

Figure 8 shows the die voltage as obtained from the time-domain EM simulation for the two packages, using identical die, decoupling capacitor, and power supply partial circuits. In each case, the voltages across the 25 independent current sources were averaged at each point in time. The results show the significant improvement of the power delivery performance that BBUL can provide. Based on the values in Fig. 8 of the first voltage minimum occurring between zero and 5 ns (the "first droop") and using Eq. (6), we can quantify this improvement in terms of a few simple numbers.

The magnitude of the first droop is found to be reduced by 36%. Whereas the loop inductance L_p of the standard package based on the simulations here is calculated to be 5.4 pH, the BBUL package exhibits an L_p of 1.4 pH; this is a reduction by 74%. Evidently, the loop inductance of the BBUL package is dominated by the lumped parasitic inductance L_{p2} of the discrete decoupling capacitors. In terms of the package inductance L_{p1} we have a reduction from 4 pH to 0.1 pH, or by 98%. This reduction is even more dramatic than one would expect from the mere reduction of the overall thickness of the package by 90%, highlighting the importance of taking into account the details of the package design.

The results described above show a clear benefit of BBUL for power delivery. Further improvements could apparently be made. For example, the closeness of the decoupling capacitors to the die makes them highly effective locally. Thus, an optimization of their placement may be interesting. Furthermore, a reduction of the nominal value of the decoupling capacitance can be investigated (under careful consideration of not only the first droop as in this paper). Physically smaller capacitors have smaller parasitic inductance, leading to further reduction of the package loop inductance in BBUL. In addition, choosing smaller capacitors yields increased flexibility for placement optimization and cost reduction.

4. Thermal modeling

Thermal management does not only face the challenges of increasing power and power density but also the increasing non-uniformity of the power distribution. Studies have shown that the thermal performance of a package is significantly affected by the non-uniformity of power dissipation [9]. The transistor junction to case thermal resistance with non-uniform power distribution might be 2 to 3 times larger than that for uniform dissipation. In order to simulate a realistic device, we randomly generated a non-uniform power map with a total power of 120 W. The average power density on the 1.19 by 1.19 cm die is 84.7 W/cm^2, and the peak power density is 5 times the average. In our model, a heat flux with this power distribution is applied on the front side of the die in order to represent the heat generated by the transistors and interconnect structures.

We studied the thermal performance of BBUL and standard flip-chip Controlled Collapsed Chip Connection (C4) packages using finite element method based commercial software, ANSYS. Cross-sectional views of the three-dimensional thermal models of BBUL and a C4 package are shown in Fig. 9. The DCAP region, as well as the core (see Fig. 1), are treated as voids and have zero thermal conductivity. The geometric parameters and thermal properties used in our models are given in Table 2. Because traditional thermal solutions are integrated at the exposed backside of the die, these thermal solutions [inte-

	BBUL	C4
T_j [°C]	98.7	97.4
T_c [°C]	78.9	79.1
Θ_{jc} [°C/W]	0.165	0.153
$\Theta_{jc,u}$ [°C/W	0.0866	0.0858
$\Theta_{jc}/\Theta_{jc,u}$	1.91	1.78
Θ_{ca} [°C/W]	0.283	0.284
$\Theta_{ca,u}$ [°C/W]	0.278	0.278

Table 3: Thermal performance comparison between BBUL and the C4 package.

grated heat spreader (IHS), heat sink, etc.] are identically utilized for both the BBUL package and the standard C4 package. To reflect the pursuits of thermal enhancement by the industry, several non-standard assumptions were made concerning the material properties and the boundary conditions in our simulation. For example, both thermal interface materials (TIM) are assumed to be 200 μm thick solder. The thermal conductivity of TIM1 is better than the TIM2 as shown in Table 3. A 30% performance gain of the heat sink is adopted to represent the numerous ongoing activities in the development of high performance heat sinks, such as heat pipe heat sink, among others [10]. While these solutions might not necessarily advance to the final products, it is evident that better thermal solutions are essential for the thermal budget of the future processors. In our model, a slightly better than natural convection is used on the surface of the printed circuit board, and the ambient temperature is assumed to be 45 °C. The junction to case thermal resistance Θ_{jc} and the case to ambient thermal resistance Θ_{ca} are typically used to characterize the overall thermal performance. The definitions are

$$\Theta_{jc} = \frac{T_j - T_c}{Q}, \quad \Theta_{ca} = \frac{T_c - T_a}{Q} \tag{8}$$

where Q is the total power, and T_a, T_j, and T_c are the ambient temperature, the maximum junction temperature, and the temperature at the center of the IHS, respectively. Traditionally, chip manufacturers are responsible for the Θ_{jc} and original equipment manufacturers (OEMs) or heat sink vendors provide the solution for Θ_{ca}. The two thermal resistances might not be easily separable as the power density keeps increasing. Closer cooperation between chip manufacturers and vendors is needed to address the problem at both chip level and system level.

The temperature contour maps of both packages on the front side of the die are shown in Fig. 10. The summary of results is given in Table 3. $\Theta_{jc,u}$ and $\Theta_{ca,u}$ are the junction to case and the case to ambient thermal resistances of the package with uniform heating. The maximum junction temperatures are 98.7 °C and 97.4 °C for BBUL and the C4 package, respectively, and they both appear at the hot spot with peak power density. The C4 package has slightly better thermal performance (2.5%) because the thicker ball grid array (BGA) layer can spread heat better

(a) BBUL package.

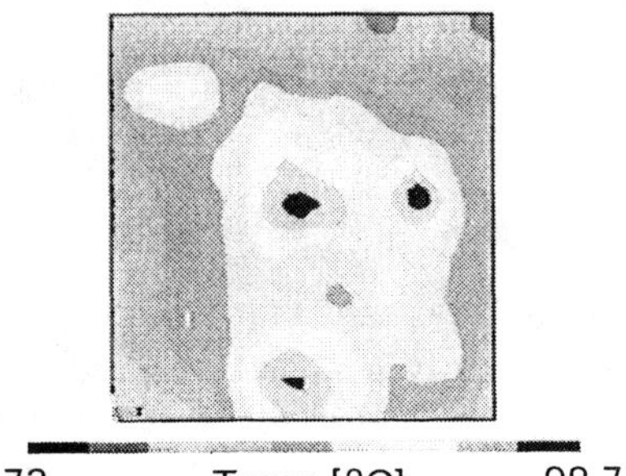

(b) C4 package.

Figure 10: Temperature contour maps on the die of the packages.

than the thin build-up layer when the DCAP regions are treated as voids in our models. From Table 3 we can also see that the Θ_{ca} of both packages with non-uniform power map is close to the packages with uniform heating, which indicates that the heat spreading in IHS is adequate. The $\Theta_{jc}/\Theta_{jc,u}$ ratio is dependent on power map and less than 2 for both packages with the power map used. Overall, the simulation shows that the thermal performance of the BBUL package is similar to that of the C4 package. As less than 5% of the thermal energy is removed through the front side of the silicon die, switching to BBUL has a minor impact on the thermal performance.

5. Mechanical attributes

Temperature excursions always occur during microprocessor use due to the turning on and off of devices. Materials with differing thermal expansion properties in intimate contact with one another will necessarily impart forces on each other. It can be expected that the BBUL package will impose relatively small forces on the surface of the die, as compared to a typical flip-chip package, during such temperature excursions. The reason for this expectation is that the relative thinness of the build-up layers, which are the only structures in contact with the die surface, means that forces which are small in magnitude will generate large deformations in the layers. Preliminary finite element modeling has shown that the shear stresses and shear plastic strains caused in vias on the die surface are smaller when using BBUL than for a flip-chip package, in some cases by more than a factor of two. However, it is possible that differences in stress concentrations due to the two packaging schemes will be different, and this remains to be investigated. A detailed discussion of the mechanical analysis of a BBUL package will be presented elsewhere.

6. Conclusions

We have described the BBUL packaging technology and some of its conceptual, process, routing, electrical, thermal and mechanical characteristics. Clear benefits for the problems of escape routing and power delivery have been identified and demonstrated. Thermally, the BBUL package performs similarly to existing technologies but is expected to be compatible with innovative thermal solutions. Furthermore, BBUL should impose a reduced level of thermomechanical stresses onto the active die surface and, thus, support the introduction of on-die dielectrics that are sensitive to such stresses.

Acknowledgments

The authors would like to thank K. Radhakrishnan, R. Nair, Y. L. Li, and A. Watwe of Intel's Assembly Technology Development division and R. Chen of Sigrity, Inc., for useful discussions.

References

[1] Hasan, A., Sarangi, A., Baldwin, C. S., Sankman, R. L., and Taylor, G. F., 2000, "High performance package designs for a 1 GHz microprocessor," *Proc. IEEE Electronic Components and Technol. Conf. (ECTC)*, Las Vegas, May 21–24, pp. 1178–1184.

[2] Lau, J. H., 2000, "Critical issues of wafer level chip scale package (WLCSP) with emphasis on cost analysis and solder joint reliability," *Proc. IEEE/CPMT Int. Electronics Manufact. Technol. Symp. (IEMT)*, Santa Clara, Oct. 2–3, pp. 33–46.

[3] Kawahara, T., 2000, "SuperCSP™," *IEEE Trans. Adv. Packag.*, **23**, pp. 215–219.

[4] Solberg, V., Light, D., and Fjelstad, J., 2000, "Reliable and low cost wafer level packaging: process description and qualification testing results for wide area vertical expansion (WAVE™) package technology," *Proc. IEEE/CPMT Int. Electronics Manufact. Technol. Symp. (IEMT)*, Santa Clara, Oct. 2–3, pp. 108–114.

[5] Yang, H., Elenius, P., and Barrett, S., 2000, "Ultra CSP™ bump on polymer structure," *Proc. Int. Symp. Adv. Packag. Mater.: Processes, Properties and Interfaces*, Braselton, Mar. 6–8, pp. 211–215.

[6] Tummala, R. R., Rymaszewski, E. J., and Klopfenstein, A. G., eds., 1997, *Microelectronics Packaging Handbook*, Chapman & Hall, New York, 2nd edition.

[7] Li, Y. L., Figueroa, D. G., Chickamenahalli, S. A., Chung, C. Y., Yew, T. G., Cornelius, M. D., and Do, H. T., 2000, "Enhancing power distribution system through 3D integrated models, optimized designs, and switching VRM model," *Proc. IEEE Electronic Components and Technol. Conf. (ECTC)*, Las Vegas, May 21–24, pp. 272–277.

[8] Kodali, V. P., 2001, *Engineering Electromagnetic Compatibility: Principles, Measurements, Technologies, and Computer Models*, IEEE Press and Wiley, New York, 2nd edition.

[9] Watwe, A., and Viswanath, R., 1999, "Localized power concentration on die: thermal impact to processor performance and package design," *Intel Assembly and Test Technol. J.* (Intel internal), **2**, pp. 139–155.

[10] Garner, S. D., and Toth, J. E., 1997, "Heat pipes: a practical and cost effective method for maximizing heat sink effectiveness," *Proc. Pacific Rim/ASME Int. Intersociety Electronic and Photonic Packag. Conf. (InterPACK)*, Kohala Coast, June 15–19, Vol. 2, pp. 1897–1902.

Proceedings of
2001 ASME International Mechanical Engineering Congress and Exposition
November 11–16, 2001, New York, NY
EPP-Vol. 1

IMECE2001/EPP-24704

ALTERNATIVE COMPLIANT INTERCONNECTS – THERMO-MECHANICAL RELIABILITY, DESIGN AND TESTING

Lunyu Ma, Qi Zhu, and Suresh Sitaraman
Computer-Aided Simulation of Packaging Reliability (CASPaR) Lab
The George W. Woodruff School of Mechanical Engineering
Georgia Institute of Technology
Atlanta, GA 30332-0405
Voice: 404-894-3405
FAX: 404-894-9342
suresh.sitaraman@me.gatech.edu

ABSTRACT

Two types of novel compliant interconnects are being proposed in this paper. The advantages of these two types of compliant interconnects are: accommodation of the mismatch due to different thermal expansions in electronic packages, integration with wafer-level fabrication process, low cost, fine pitch and high I/O density. The first type is a highly-compliant cantilevered spring interconnect for the next generation packaging and probing technology. To understand the reliability of the package with this novel compliant spring interconnect structure and the typical behavior of sliding contact, test vehicles with different orientations of the cantilevered springs (21 μm pitch) have been fabricated, assembled and subjected to thermal cycling test. In-situ resistance and temperature measurements have been conducted. The second type is a compliant free-standing interconnect, One-Turn Helix (OTH) structure. This structure is built as an one-turn strip helix in order to get good compliance in space. It can be fabricated through conventional photolithography-based IC fabrication process. Optimal design parameters have been identified for this structure taking into consideration the thermo-mechanical and electrical behavior of this unique structure.

INTRODUCTION

The miniaturization of devices, high I/O density, fast clock speed, good functionality and system integrity, low cost and good reliability are always the driving forces of the electronic industry. The rapid advances in IC design and fabrication continue to challenge and push the electronic packaging technology, in terms of size, performance, cost, and reliability. The minimum feature size in IC components has reached the scale of about 180 nanometers, requiring chip-to-substrate interconnect pitch in the range of few microns. There are not many cost-effective and manufacturable solutions today to be used as the chip-to-substrate interconnection in such a fine pitch, and to be able to accommodate the coefficient of thermal expansion (CTE) mismatch among different components without sacrificing reliability and performance. This has become a bottle-neck seriously limiting the advances in the electronics industry.

In this paper, two types of novel compliant interconnects have been discussed. The first one is a novel highly-compliant cantilevered spring interconnect structure, developed and fabricated by research teams at Georgia Tech, Xerox PARC, and Nanonexus Inc. [Haemer, et al., 2000] based on the classic works of Hoffman and Thornton [e.g. Hoffman and Thornton, 1982]. This new structure shows us a possible direction of interconnection technology development with great promises and lots of potential benefits. It can satisfy the demands for increasing system miniaturization and functionality, lower cost and higher efficiency from the electronic packaging industry. The minimum pitch size of this highly compliant cantilevered spring structure has reached as small as 6 μm, which has already gone far beyond the fine-pitch requirements of International Technology Roadmap for Semiconductor (ITRS) for year 2012 [ITRS, 1999]. Also another unique advantage of this technology is that it enables the no-soldering contact between the spring and the bonding pad. The second type of compliant interconnect is called 'One-Turn Helix' (OTH) structure. It looks like a spiral spring to obtain the good multi-directional compliance to accommodate the probing or assembly force and CTE mismatch during the thermal cycling. It is based on conventional IC fabrication technology. The fabrication of both types of interconnect can be integrated into wafer-level processing as a batch process, and therefore, the cost can be reduced, especially when I/O count is high. Due to its high compliance, both structures can accommodate more differential displacement due to the CTE mismatch of different materials in electronic packages than conventional interconnect structures.

NOMENCLATURE

Compliant Interconnect, OTH, CTE, Thermal Cycling, Reliability.

1. COMPLIANT CANTILEVERED SPRING

1.1. FABRICATION

As shown in Hoffman and Thornton's work [Hoffman and Thornton, 1982], the intrinsic stress can be introduced by controlling the Argon pressure during DC magnetron sputtering process. A compressive-to-tensile stress gradient is built up in the thin spring metal layer during DC sputtering. The fabrication process of the compliant, cantilevered spring interconnect can be summarized as Figure 1.

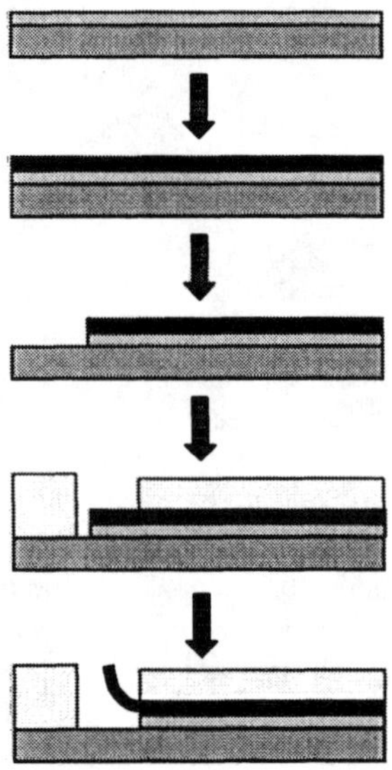

Figure 1 The Fabrication of The Compliant Cantilevered Spring Structure

The released spring structure is shown in the Figure 2. In this figure, the pitch of the spring is about 6 μm. These compliant interconnect structures are capable of satisfying ultra fine-pitch requirements. As the spring fabrication uses standard deposition and lithography processes of IC fabrication technique, the technology is expected to be low cost, high-yielding and can be integrated into wafer level processes.

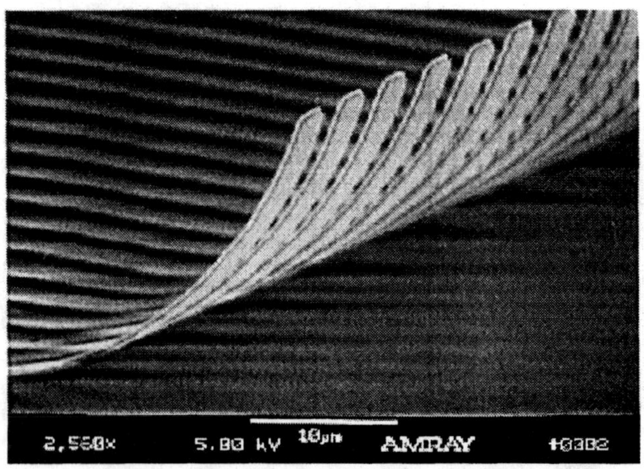

Figure 2 SEM (Scanning Electron Micrograph) Picture of an Array of Springs

1.2 ASSEMBLY OF TEST VEHICLE AND THERMAL CYCLING EXPERIMENTS

The test vehicles were fabricated by sputtering and patterning spring metal on quartz substrate and assembling on glass or silicon die with bonding pads. The accuracy of alignment is required to be about 1 μm due to the fine feature size. Therefore optical alignment system was used in test vehicle assembly. Furthermore, because this highly compliant interconnect technology is planned to be used in some proprietary application, the underfill material and one of substrate and die shall be transparent. Besides that, all the assembly processes shall be conducted at room temperature. Therefore, a commercial transparent underfill material is selected for test vehicle, which is curable with UV (ultraviolet) light at room temperature.

The daisy chain structure is integrated into the test vehicle and used to measure the in-situ resistance change vs. temperature. The daisy chain is formed between the bonding pads on the die and the springs at the end of metal traces on the substrate, as shown in Figure 3. The spring is about 17 μm wide and the pitch size is about 21 μm. Two different orientations of the compliant spring are fabricated in the test vehicle, as shown in the Figure 4. In the horizontal direction, there are 11 devices in a sequence, containing over 700 springs. These 11 horizontal devices are located on the horizontal neutral axis. Therefore during the thermal cycling, there will be no differential thermal expansion in the '*Y*' direction. All differential thermal expansions will occur in the horizontal or '*X*' direction. Another 6 devices are located in the vertical direction, which form independent daisy chains. For the vertical rows of springs, on the other hand, both horizontal and vertical thermal expansions will influence the test results due to the offsets from the horizontal and vertical neutral axes.

The test vehicle is subjected to the thermal cycling test in a pogo cassette. The thermal cycling test is based on JEDEC Temperature Cycling Standard (JESD22-A104-A). For regular office type of electronic equipment, experiments from room temperature to 125°C seem more appropriate [Syed and Doty, 1999]. In this preliminary work, room temperature-to-125°C experiments are done; the –55°C-to-125°C experiments are currently underway. During each cycle, the temperature is first ramped from room temperature to 125°C at a ramping rate of 20°C per minute. Then the temperature is dwelled at 125°C for 10 minutes. After that, the temperature is lowered to room temperature at a rate of 20°C per minute. Finally, the temperature is dwelled at room temperature for 10 minutes. So the total time for one cycle is about 30 minutes. During the experiments, the sampling is done every 0.1 minute. So far, 15 test vehicle samples have been assembled and subjected to thermal cycling experiments.

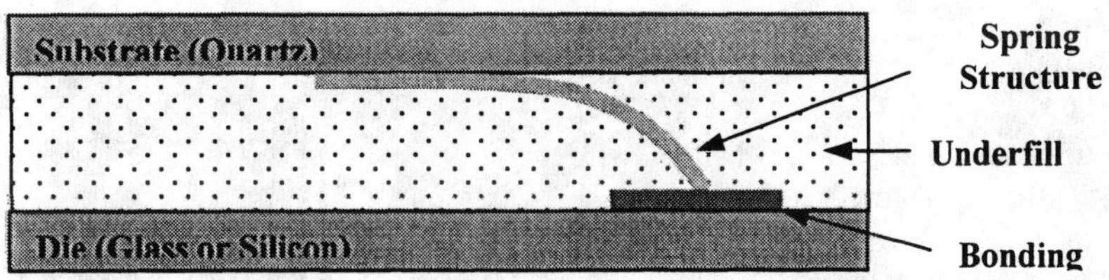

Figure 3 Schematic of Test Vehicle Cross Section

As mentioned earlier, the assembly of test vehicle is conducted using an optical aligner. The substrate is fixed on a stationary stage by vacuum sealing. The die is attached to the mobile stage also by vacuum sealing. The mobile stage has four degrees of freedom, '*X*', '*Y*', '*Z*' and 'θ_z'. Although the spring is highly compliant, it is still very fragile under external impacts, such as shock from suddenly rapid movement in viscous uncured underfill. So a coarse alignment is made before applying underfill. After that, the underfill material is dispensed onto the die. The die with bonding pads on top surface is moved up towards the substrate with springs facing down. The spring tips penetrate the underfill and get in contact with pads of the die. At the same time, the upward movement of the die also helps the formation of a thin, uniform layer of underfill material between substrate and die. Before the springs are fully compressed, a fine alignment is made to ensure accurate positioning. Finally the springs are compressed further to get good contact between tips and pads. To prevent the misalignment and constrain the spring tips onto the bonding pads, the test vehicle is immediately exposed under UV light to cure the underfill material.

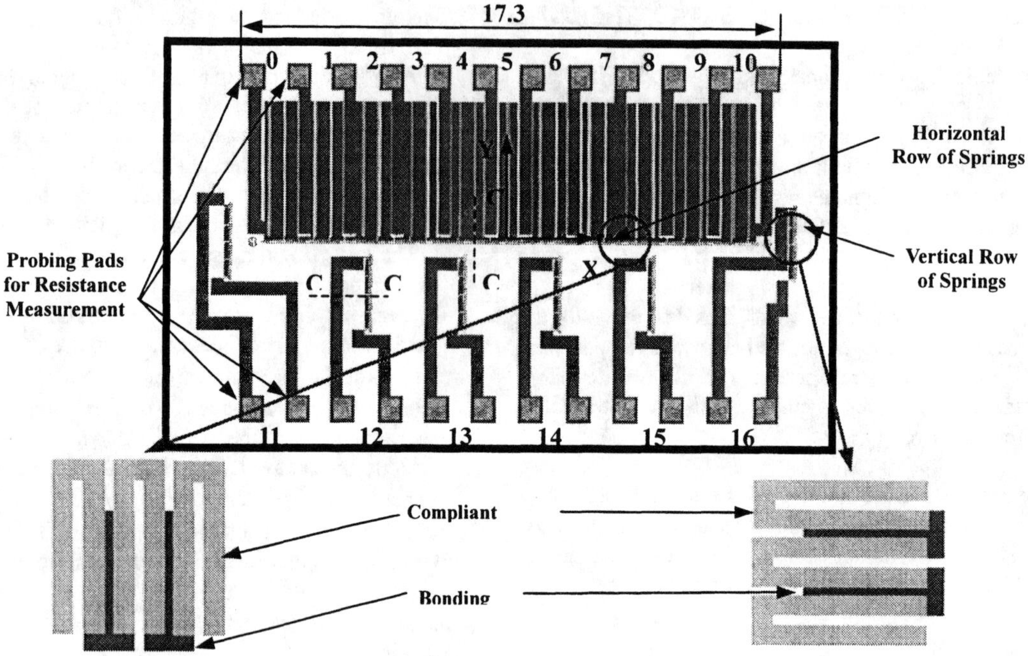

Figure 4 Layout of Test Vehicle and Zoom-out

1.3 RESULTS AND DISCUSSIONS

From the thermal cycling test, several phenomena are found:

1) It is found that the resistances (R) change with temperature (T). As seen in Figure 5, the change of resistance (ΔR) exactly follows the change of temperature (ΔT). The measured resistance change with temperature is approximated as:

$$\frac{\Delta R}{R \cdot \Delta T} \approx 1.2 \times 10^{-3} \quad 1/K \qquad (1)$$

Where:

R – the resistance at room temperature in the beginning of experiment

ΔR – the change of resistance

ΔT – the temperature range during one thermal cycle

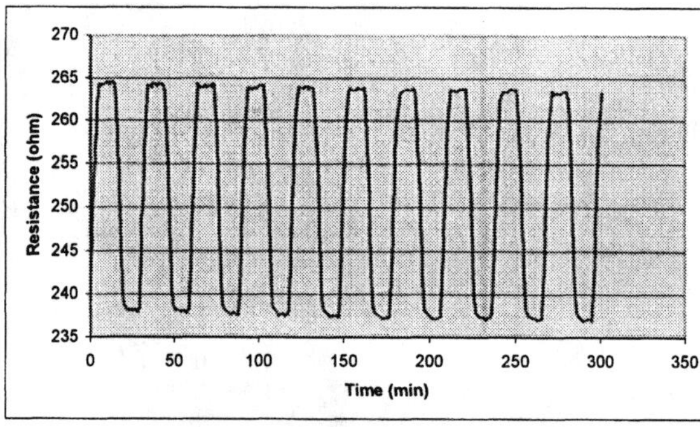

(a)

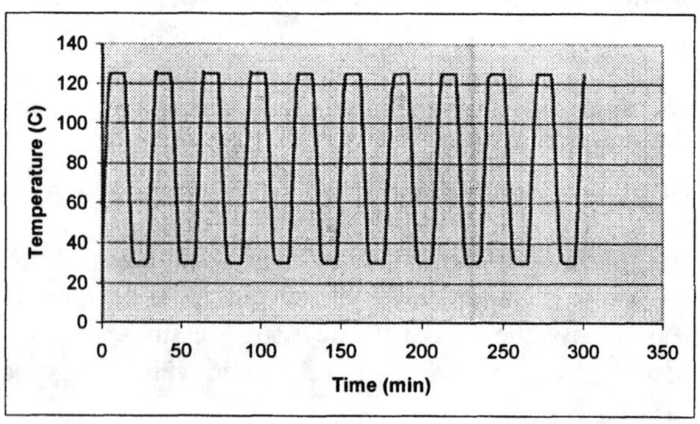

(b)

Figure 5 Resistance Change with Temperature (a) Resistance vs. Time; (b) Temperature vs. Time

2) The "burn-in" phenomenon [Ma, et al., 2001] is found for both the horizontal and vertical devices during the thermal cycling test, as shown in Figure 6. The "burn-in" is continuous and irreversible in the thermal cycling test, even if there is interruption.

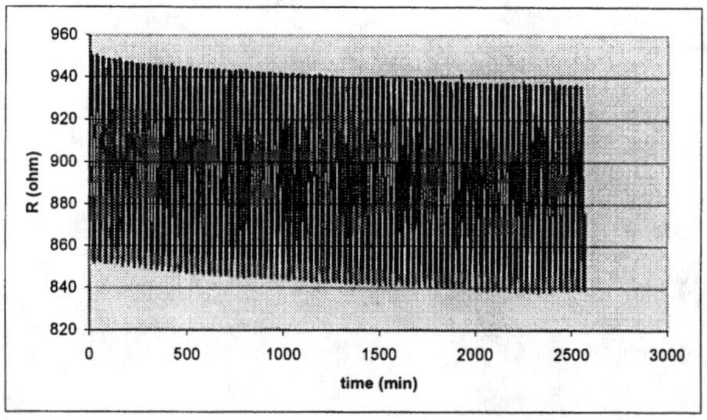

(a)

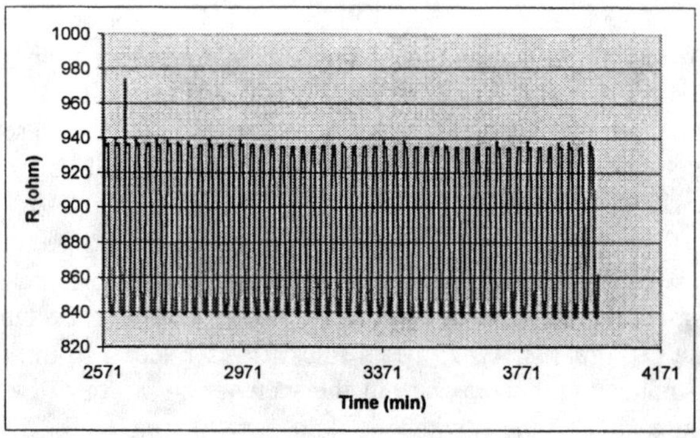

(b)

Figure 6 Resistance vs. Time of an Arrays: (a) In the First 80 Cycles; (b) In the Next 40 Cycles

(3) The experiment shows that the reliability of the horizontal device is much better that of the vertical devices. For daisy chains in the vertical devices, the reliability is greatly dependent on the distance from the vertical neutral axis in the horizontal direction. This can be seen in Figure 7. Device 14, 15 and 16 are three devices in the vertical arrays on the right side of neutral axis in the same test vehicle, as seen in Figure 2. Device 14 is the closest to the vertical neutral axis on the test vehicle. Device 16 is the farthest from the neutral axis. Device 15 is located between 14 and 16. The test vehicle is subjected to 500 minutes of thermal cycling. As seen in Figure 7(c), device 16 fails first, almost right at the beginning of test. Device 15 fails after about 80 minutes, and device 14 survives throughout 500 minutes. This shows that devices close to neutral axis have better reliability than other devices due to less severe differential displacement of CTE mismatch.

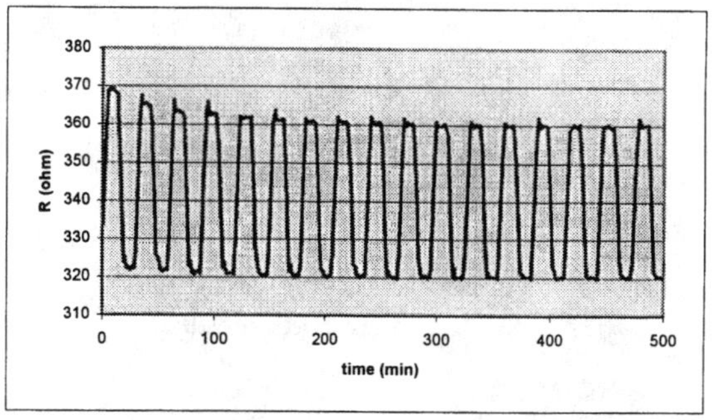

(a) Device 14

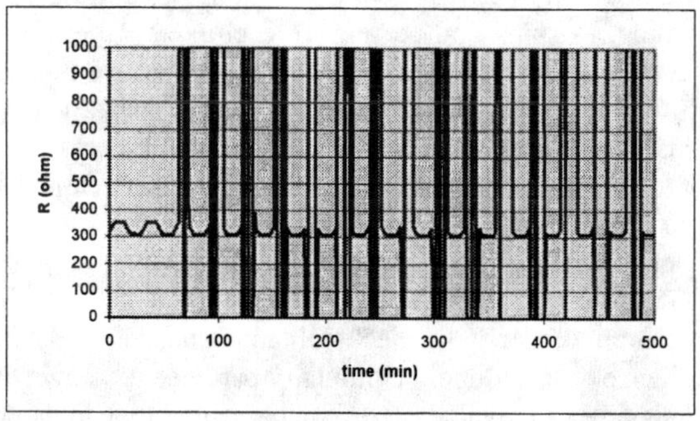

(b) Device 15

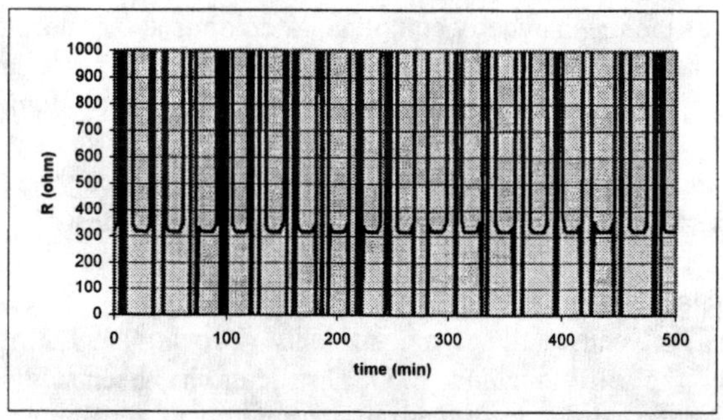

(c) Device 16

Figure 7 Resistance Measurement for Three Adjacent Vertical Arrays

2. COMPLIANT ONE-TURN HELIX (OTH)

The second proposed novel interconnect structure is called One-Turn Helix (OTH) structure. It looks like a spiral spring with good compliance in all three directions, as seen in Figure 8.

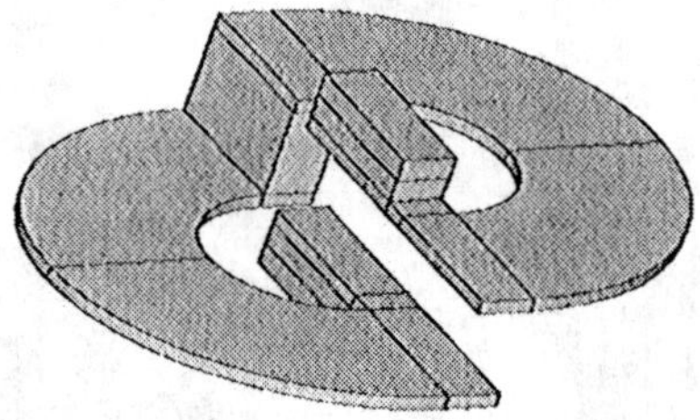

Figure 8 OTH 3D Structure

2.1 FABRICATION PROCESSES

This structure can be built using conventional photolithography based IC fabrication techniques. After depositing, masking and etching for each metal layer, the 3-D interconnects are fabricated through photosensitive polymeric layers. Then the polymeric layers are finally etched away, the free-standing metal interconnect structure can be created. The feasible fabrication process steps have been developed, and will be experimented at Georgia Tech.

2.2 GEOMETRY DESIGN

A parametric study on electrical inductance and mechanical compliance was done. Optimum parameters have been recommended to reduce electrical parasitics and to improve mechanical compliance.

2.2.1 STUDY ON ELECTRICAL INDUCTANCE

For the mechanically compliant interconnect, it is necessary to ensure that the electrical parasitics of such type of structure comply with the acceptable design limits. The essential goal of the electrical design is to maintain signal integrity and fidelity in signal path. R (resistance), L (inductance) and C (capacitance) are often used to qualitatively describe the electrical performance of interconnects. Among them, self-inductance is the most critical factor in determining the delta-I noise that causes the signal infidelity [Brown, 1999]. In this paper, the self-inductance of OTH structure has been studied.

The first order analytical approximation of self-inductance is useful for quick estimation and trade-off between various design parameters in the preliminary design. In the absence of magnetic materials, the self-inductance of a rectangular-strip helix under low frequency condition can be reasonably approximated as [Grover, 1980]:

$$L_{self} = 0.001F(2R)N^2 - 0.004\pi RN(g+h) \qquad (2)$$

Where:

N = number of turns =1(in this work)
H = distance between adjacent wires
W, T = width and thickness of helix beam
R = mean radius of helix
B = equivalent length = $N \times H$
F = table listed value corresponding to $B/2R$
g, h = correction terms corresponding to H, W, T and N

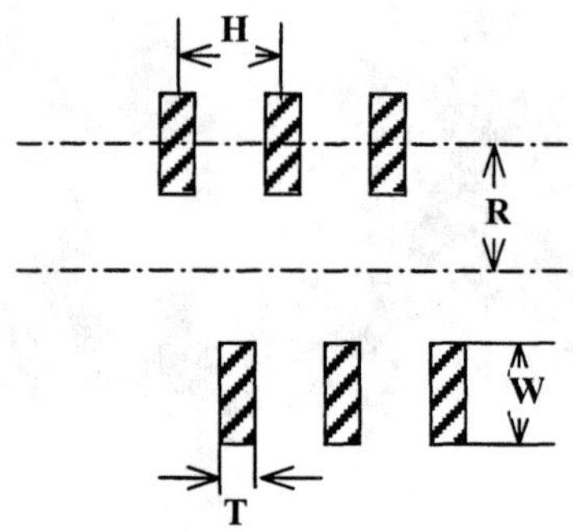

Figure 9 Geometric Parameters Used in the Inductance Calculation

Some preliminary studies have been done based on the analytical solution (Equation 2). When structure vertical height and beam cross section dimension are kept the same, the self-inductance will increase significantly from 0.007 nH to 0.33 nH with the increase of mean radius of helix from 10 to 100 μm (Figure 10). If we increase the vertical height and keep the other parameters the same, the inductance will also increase. This means that taller the structure, higher will be inductance value (Figure 11). When beam thickness is fixed at 3 μm, the self-inductance decreases with the increase of width of beam (Figure 12). When width is kept at 30 μm, inductance decreases with the increase of T (Figure 13). The dependency of inductance on vertical height, beam thickness and width is much less compared to the dependency of inductance on the mean radius of helix.

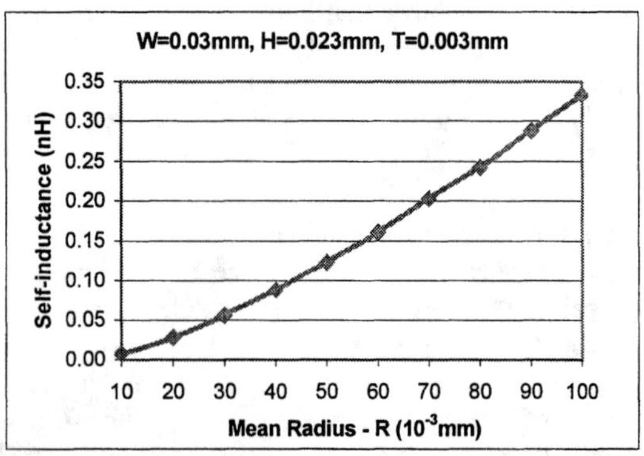

Figure 10 Effect of the Mean Radius on the Inductance

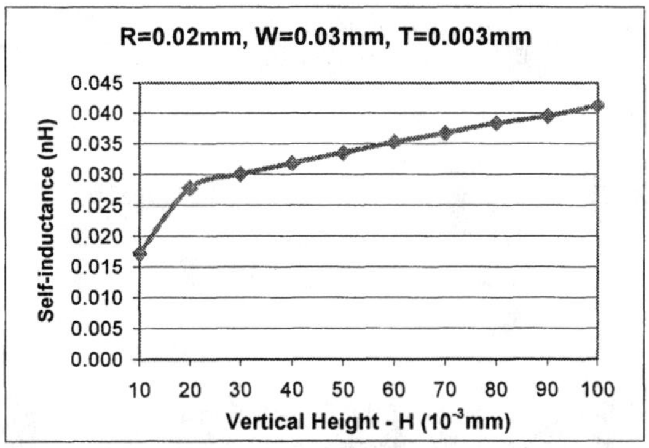

Figure 11 Effect of the Vertical Height on the Inductance

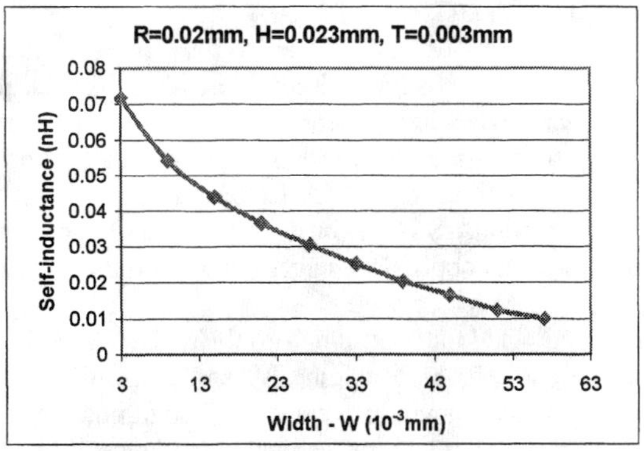

Figure 12 Effect of the Width on the Inductance

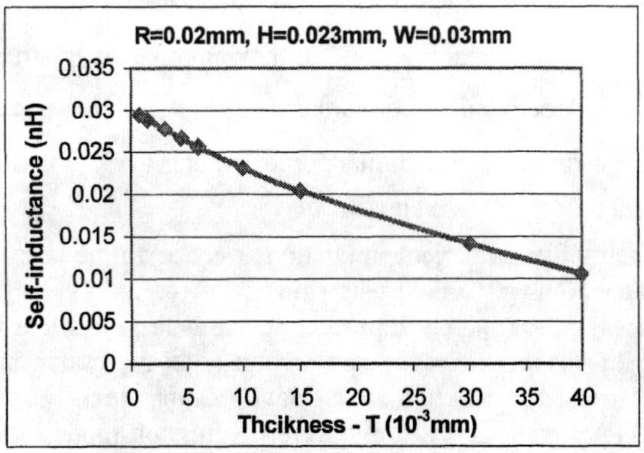

Figure 13 Effect of the Thickness on the Inductance

From the electrical design point of view, the self-inductance should be as low as possible. Based on the above studies, it is seen that the structure should be thicker and wider, and with smaller radius and lower vertical height.

2.2.2 STUDY ON MECHANICAL COMPLIANCE

Mechanical compliance is also a very important design parameter for the compliant OTH interconnect. In this work, we study the effects of geometric parameters on the compliances in all three directions using numerical simulation.

A finite element model has been developed to understand the deformation behavior of a single OTH structure under the mechanical loading. Through the model, the relationship between the probing or contacting force versus the displacements can be determined. Copper, the compliant interconnect material, is modeled as temperature independent, bi-linear kinematic hardening material. The numerical simulations were conducted by applying forces in the *x*, *y* and *z* directions orthogonally. The resulting displacements u_x, u_y and u_z were determined and the directional compliance, C_x $(= u_x/F_x)$, C_y $(= u_y/F_y)$, C_z $(= u_z/F_z)$, were obtained.

(A) EFFECT OF MEAN RADIUS OF HELIX (R)

The mean radius of helix has a major effect on the compliance, especially in the vertical direction. In this simulation, the vertical height, helix thickness and width are kept at 20, 10 and 20 μm respectively. The mean radius of helix is varied form 5 to 50 μm. As expected, the simulation results show that the compliance increases with the increase of helix radius, as shown in Figure 14. The compliance increase in the vertical direction (*y*) is more prominent than those in the other two directions.

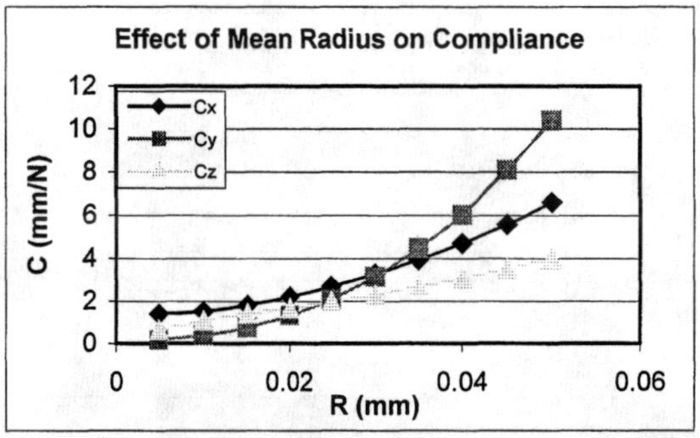

Figure 14 Effect of the Mean Radius on the Directional Compliance

(B) EFFECT OF VERTICAL HEIGHT (H)

The vertical height is varied form 5 μm to 50 μm, but other geometric parameters are kept the same. The width and thickness of helix strip are 20 μm and 10μm. The mean radius of helix is 30 μm. Figure 15 shows the effects of vertical height to the three directional compliance. The compliance in *x* and *z* directions increase with the vertical height. However, the compliance in *y* direction does not change much.

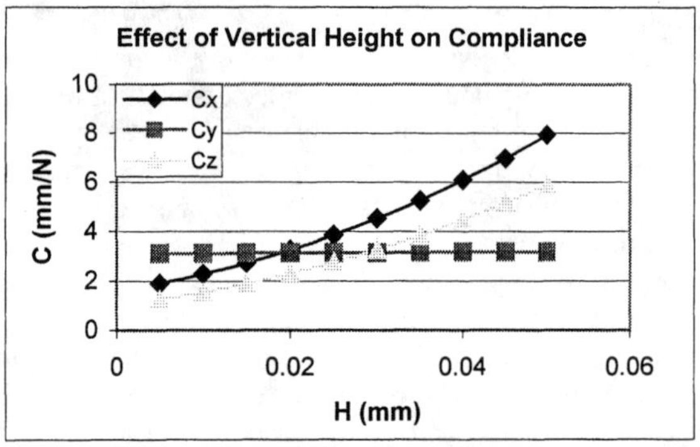

Figure 15 Effect of the Vertical Height on the Directional Compliance

(C) EFFECT OF THICKNESS (T)

The thickness of strip helix is simulated from 3μm to 30μm. The width and height of strip helix are kept as 20 μm and the mean radius of helix is 30 μm. Figure 16 shows the effect of T to structure compliance. The thinner beam has better compliance in all directions than the thicker beam, especially in y direction. It can be seen that when beam is thick, the compliance is quite low and almost the same in all three directions. The increase in vertical directional compliance is evident when structure is thin.

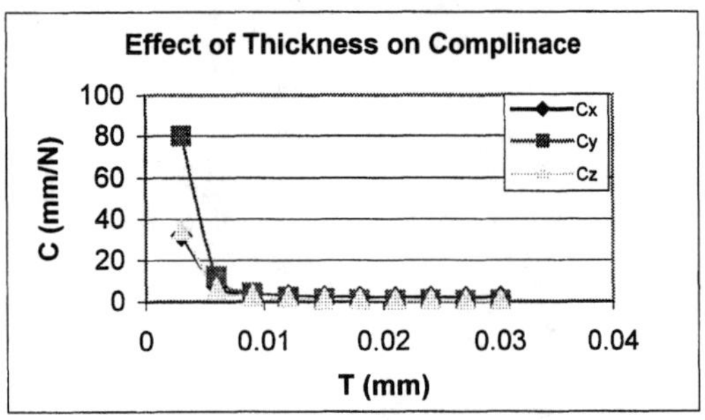

Figure 16 Effect of the Thickness on the Directional Compliance

(D) EFFECT OF WIDTH (W)

When the thickness, radius and height are kept at 10 μm, 30μm and 20 μm, the width is changed from 5 to 50 μm, Figure 17 shows the effect of helix width in three directional compliance. It is obviously that the narrower beam, the better compliance.

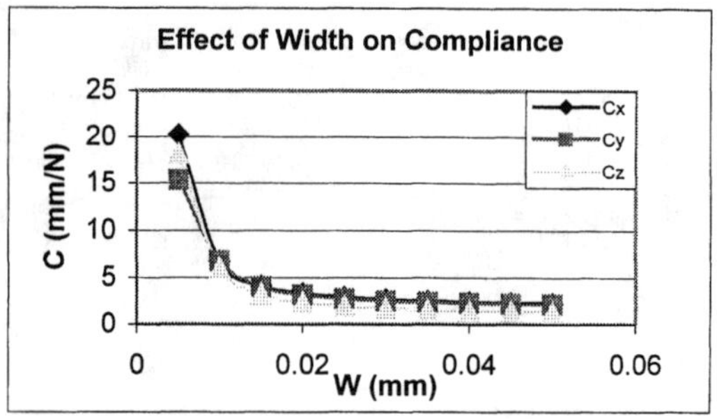

Figure 17 Effect of the Width on the Directional Compliance

In summary, thinner and narrower beam with larger radius and vertical height will have better mechanical compliance.

2.2.3 OPTIMIZATION

From the previous parametric studies on the electrical inductance and the mechanical compliance of OTH structure, it is known that when a certain geometric parameter is increased to improve the mechanical compliance, the self-inductance and stresses will also be increased under the same loading condition. So response surface and optimization methodology was applied to select among these design trade-offs.

In this work, four design variables, mean radius *(R)*, vertical height *(H)*, helix width *(W)* and helix thickness *(T)*, with three different levels are selected. The ultimate goal is to improve three-directional mechanical compliance, decrease the self-inductance and ensure the structure will work without yielding. Seven response variables, compliance in *x* direction (C_x), compliance in *y* direction (C_y), compliance in *z* direction (C_z), maximum stress in structure due to *x* directional force loading ($\sigma_x^{\max}$), maximum stress in structure due to *y* directional force loading ($\sigma_y^{\max}$), maximum stress in structure due to *z* directional force loading ($\sigma_z^{\max}$) and self-inductance (L_{self}), are chosen.

Twenty five numerical simulations were performed based on the Face Center Cube Design matrix [Myers, 1995]. Seven surfaces approximately representing the data points had been built. In this work, a complete second-order regression model was used to fit nonlinear data. Commercial software MINITAB was used to build seven regression models in this work.

The DSIDES, developed at Georgia Tech [SRL, Georgia Tech], has been used to compromise all the design trade-offs in mechanical compliance, stress and electrical inductance. We set the target of each response as: $C_x = C_y = C_z = 30\,{}^{mm}\!/_{N}$; $\sigma_x^{\max} = \sigma_y^{\max} = \sigma_z^{\max} = 100MPa$ and $L_{self} = 0.01nH$. Also the constraints were set as:

$\sigma_x^{max} \le 173MPa,\ \sigma_y^{max} \le 173MPa,\ \sigma_z^{max} \le 173MPa$.

The reasons of selection of those values as targets and constrains are: typically the compliance of flip chip interconnects will not exceed 30mm/N; 0.01nH is the typical self-inductance value for solder joint which has the shortest signal path; the yield stress of copper material is about 173 MPa and typically copper structure can work under elastic region when stress is about 100 MPa.

Based on above targets and constraints, the optimum parameters are: $R = 29\mu m, H = 28\mu m, W = 22\mu m, T = 7\mu m$.

The corresponding compliance, stresses and inductance are:

$$C_x = 7.78\,{}^{mm}\!/_{N}, C_y = 8.86\,{}^{mm}\!/_{N}, C_z = 7.56\,{}^{mm}\!/_{N},$$

$$\sigma_x^{max} = 156.98MPa, \sigma_y^{max} = 94.09MPa, \sigma_z^{max} = 143.78MPa,$$

$$L_{self} = 0.0625nH$$

2.3 THERMO-MECHANICAL PERFORMANCE

Once an optimum geometry has been identified, the OTH was placed as a silicon to organic substrate interconnect as shown in Figure 18 and the behavior of the interconnect under thermal loading was simulated using a strip model, which is also known as Generalized Plane Deformation (GPD) model. Thermally induced stress and strain have been studied through this model.

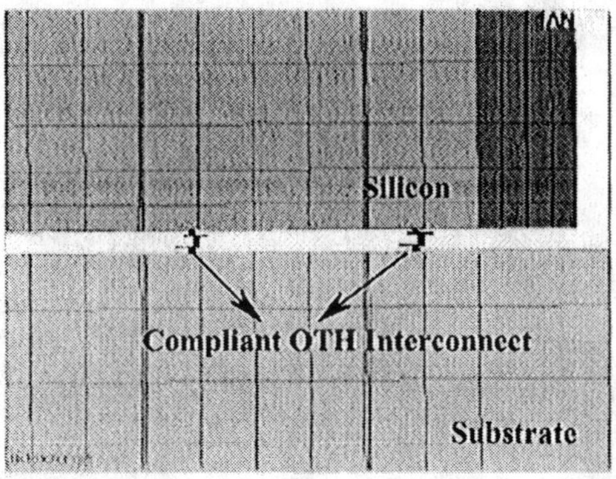

Figure 18 FEA Model of the Package Assembly with OTHs

The proposed compliant interconnects are studied for a silicon on a high CTE (16 ppm/°C) substrate. The high CTE substrate is intentionally chosen to produce a high CTE mismatch between the silicon and the substrate. In the work presented here, the compliant interconnects are assumed to be soldered to the substrate using high modulus Au-Sn solder rather than Pb-Sn eutectic solder. This is because, as shown later, the complaint OTH structure will be able to accommodate effectively the differential displacements due to CTE mismatch, and a stiffer solder is essential to make use of the compliant interconnect to ensure that solder strains are minimal. In this work, the silicon die is modeled as a modified linear isotropic material with temperature dependent elastic modulus. Au-20Sn solder is modeled as a multi-linear kinematical hardening elastic-plastic material.

The model was thermal cycled through four load steps: from reflow temperature 280°C to room temperature 25°C, from 25°C to 125°C, then to -55°C, and finally to 25°C again. This is the temperature range of MIL-STD-883 test. The die size in the model is 5mm x 5mm. The package was assembled by the 100 area arrayed compliant OTH structures with the optimized geometry.

Figure 19 shows the Von Mises stress and accumulative equivalent plastic strain distributions in the package assembly with die, substrate, complaint OTH interconnects and Au-20Sn solder. In the packaging assembly, the outermost interconnect is located about 2.3 mm from the neutral point. The outermost interconnect undergoes the maximum stress and strain. Over one cycle, the effective plastic strain range in the compliant interconnect is 0.00335. This plastic strain range would roughly translate to 1782 cycles [Engelmaier, 1987], and therefore, the compliant interconnect structure is likely to last more than 1000 cycles, typically used in industry qualification of electronic packaging assemblies. The accumulated equivalent plastic strain in the Au-20Sn solder is negligible. The FEA model with large DNP will be studied in a future publication.

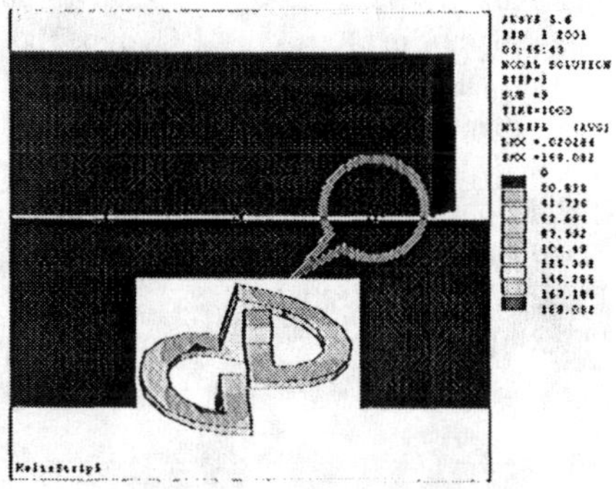

(a) Von Mises Stress

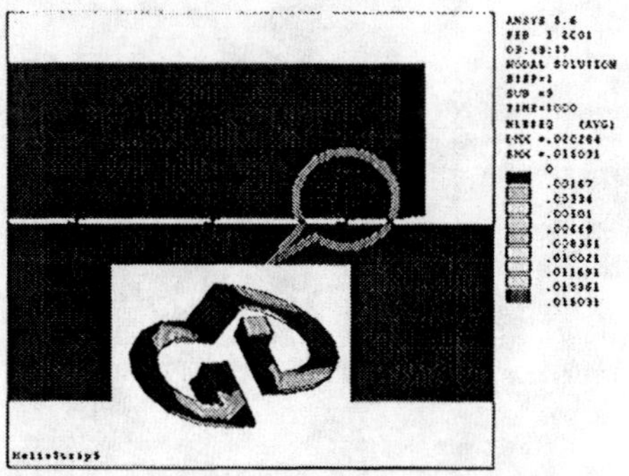

(b) Equivalent Plastic Strain

Figure 19 Von Mises Stress and Equivalent Plastic Strain in OTH

It should be pointed out that Au-20Sn is selected due to the higher modulus. In practical application, Au-20Sn is expensive and has high melting temperature. Therefore a solder material needs to be identified, which is stiff but lower melting point, less expensive than Au-20Sn, and good wetting.

3. SUMMARY

In this work, two unique types of novel compliant interconnects are discussed. The fabrication of both structures is based on IC fabrication technology, and can be integrated into wafer-level processing. Different studies are conducted in various aspects related to the reliability issues. In the study of the compliant cantilevered spring interconnect, test vehicle with such structure is subjected to the thermal cycling test Good reliability is found in certain orientation. The parametric study is performed on the OTH structure in terms of mechanical and electrical performance. Response surface methodology is applied to select optimum parameters of this structure. More reliability study and fabrication work are currently underway.

ACKNOWLEDGMENTS

The authors are thankful to NIST (ATP Award 70NANB8H4008) for funding the stress-engineered compliant interconnect part of this paper. Valuable collaboration, insight, and help from Xerox PARC and Nanonexus are acknowledged. The authors are also thankful to Interconnect Focus Center at Georgia Tech for funding the work on OTH.

REFERENCES

1. Haemer, J.M., Sitaraman, S.K., Fork, D.K., Chong, F.C., Mok, S., Smith, D.L. and Swiatowiec, F., "Flexible Micro-spring Interconnects for High Performance Probing", *Proceedings of 50th Electronic Components and Technology Conference*, 2000, pp. 1157-1163.
2. The International Technology Roadmap for Semiconductors Technology Needs, Semiconductor Industry Association, 1999 Edition, pp.231.
3. Hoffman, D.W. and Thornton, J.A., "Internal Stresses in Cr. Mo, Ta and Pt Films Deposited by Sputtering from a Planar Magnetron Source", *J.Vac. Sci. Tech.*, 20(3), pp. 355-358
4. Syed, A. and Doty, M., "Are We Over Designing for Solder Joint Reliability? Field vs. Accelerated Conditions, Realistic vs. Specified Requirements", *Proceedings of 49th Electronic Components and Technology Conference*, 1999, pp. 111-117
5. Brown, W.D., 1999, "Advanced Electronic Packaging", IEEE Press, New York
6. Engelmaier, W., 1987, " Results of IPC Copper Foil Ductility Round-Robin Study", IPC Publication No. 947
7. Grover, F.W., 1980, "Inductance Calculations Working Formulas and Tables" , D.Van Nostrand, New York
8. Myers, R.H. and Montgomery, D.C., 1995, "Response Surface Methodology: Process and Product Optimization Using Designed Experiments", John Wiley & Sons, New York
9. Systems Realization Laboratory (SRL), Georgia Institute of Technology
   Web Site: http://www.srl.gatech.edu/

Proceedings of
2001 ASME International Mechanical Engineering Congress and Exposition
November 11–16, 2001, New York, NY
EPP-Vol. 1

IMECE2001/EPP-24705

Design Features of Compliant Packages and Their Impact on Reliability

Ilyas Mohammed and Young-Gon Kim
Tessera Technologies, Inc.
3099 Orchard Dr., San Jose CA 95134

Abstract
It is well-known that the main cause of mechanical failure in electronic packages is due to the difference in the Coefficients of Thermal Expansion (CTE) of the silicon and the organic board. There are many packaging technologies that try to overcome this limitation; ranging from making curved connection pins (gull-wing leads) from the package to the board, as in the case of Thin Small Outline Package (TSOP), to using hard epoxy to rigidly adhere the die to the board as in the case of flip-chip packages. This paper illustrates a compliant packaging concept that minimizes the effect of the CTE mismatch between the silicon die and the board.

A summary of different packaging techniques that address the CTE mismatch problem is presented. From this summary, it is apparent that many of these techniques do not provide as high reliability as the compliant packages do, especially when the electrical connections from the package to the board (solder balls) are present directly under the silicon die as in the case of chip scale packages. As the compliant package isolates the effect of the silicon die from the substrate, the silicon has some motion relative to the substrate. This means that the interconnections from the silicon to the substrate must be designed to withstand this motion. Hence the design of these interconnections is key to maximizing the reliability of the compliant packages.

A detailed design and reliability analysis of compliant packages for different applications is presented. The design highlights the main parameters that have an effect on reliability of the package. Reliability simulation and analysis using finite element techniques is presented for different designs to highlight the key parameters that govern the reliability of compliant packages. Finally, reliability testing data is presented for different packages.

1.0 Introduction
Microelectronics technology has made enormous strides over the past few decades. Since the advent of large-scale integration of transistors in a single piece of silicon, the packaging requirements have become more stringent. The functionality on a silicon device has increased, hence increasing the number of interconnections to be provided by the package. At the same time, the size of the device has decreased, hence dramatically increasing the interconnection density per unit area. This implies that distance between solder balls (pitch) from the package to the board decreases for the same package size. In this paper, the effect of this phenomenon on reliability of the packages is studied. A summary of the different package-to-board connection methods is presented. A compliant packaging concept is presented next. Computational analysis results are presented that show the effect of the compliant layer between materials having different CTE. As this approach allows for relative movement between these materials, the design and reliability of silicon die to substrate interconnections becomes critical for package reliability. Two types of interconnection technologies are discussed. Finally, data from reliability testing of these compliant packages is presented and the failure features are analyzed. In conclusion, the salient features of the compliant packages along with the advantages and issues are listed.

1.1 Packaging Technologies
As the focus of this study is to look at reliability aspects of packaging, and in particular the reliability of interconnects from the package to the substrate, a brief look at the different packaging techniques would be informative. Due to various reasons, including infrastructure and cost, most of the packages are assembled and then mounted on board. There are many types of interconnections from the package to the board and three popular techniques as shown in Figure 1 are [1],

1. Pin grid array (PGA). This type of interconnection is typically used for high Input/Output (I/O) packages such as Application Specific Integrated Circuits (ASICs) and microprocessors as the I/O can be distributed over an area. The silicon die is usually attached to a ceramic substrate. This fact minimizes the CTE mismatch between the die and the substrate. The mismatch between the package and the board is absorbed by the long thin pins and hence make the package reliable. But this packaging technique is costly and is used mostly for high-end die. As this is a through-hole technology, the pitch cannot be small.

One of the main causes of failure in packages is due to the fact that the CTE of the silicon die is significantly smaller than that of the organic board. Due to this reason, stresses develop when a package is mounted on board and experiences temperature change. A compliant packaging concept is primarily concerned with ensuring reliability of the package by using a compliant material and not necessarily depend on the pins and leads shown in Figure 1, since these methods cannot be used when the package size and the interconnection pitch decrease. Consider the two components shown in Figure 2. The compliant package introduces a material in between these two to minimize the CTE mismatch problem.

2.1 Beam Analysis

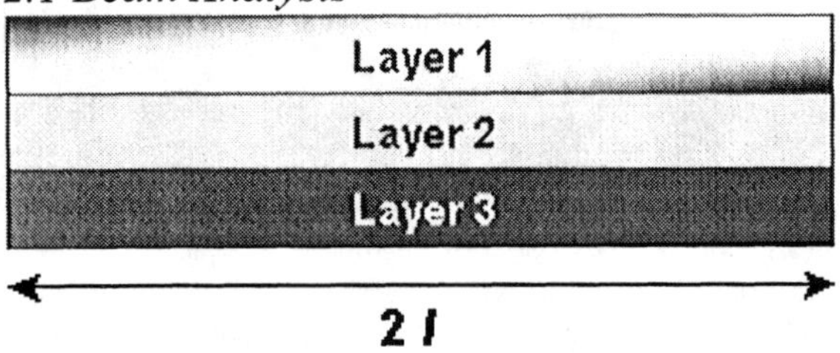

Figure 3. A tri-layer beam

The compliant layer concept can be analyzed analytically by starting with a tri-layer beam based on the original work of bi-metal strips by Timoshenko [2] and developed by Suhir [3]. Consider a beam with layers of 3 different materials as shown in Figure 3. The material in between is the potential compliant material between the silicon die and the organic substrate. Assuming the thickness is negligible compared to the radius of curvature for each layer and the shear stress in the middle layer varies linearly with thickness, the shear stress in the middle layer is given as [4],

$$\tau_{\max} = [(\alpha\,\Delta T)_1 - (\alpha\,\Delta T)_3]\frac{G_2 \tanh(\beta l)}{\beta t_2} \qquad (1)$$

where,

$$\beta = \sqrt{\frac{G_2}{t_2}\left[\frac{1}{(Et)_1} + \frac{1}{(Et)_3}\right]} \qquad (2)$$

Consider the layer 1 to be a silicon die (E_1=130 GPa, t_1=0.25 mm, α_1 = 3E-6/^{0}C) and the layer 3 to be an organic substrate (E_3=5 GPa, t_3=0.1 mm, α_2 = 20E-6/^{0}C). Assume the length of the beam is $2l$ = 10 mm. When the tri-layer beam is cooled down by 100 ^{0}C from its stress-free temperature, equation 1 can be plotted for a range of shear modulus and thickness values of the middle layer. This plot is shown in Figure 4. It can be seen from this plot that both modulus and thickness can be used to reduce the shear stress in the middle layer. If the thickness of the middle layer has to be kept at a minimum due to package thickness requirements, a compliant material is a very convenient way of reducing the shear stress.

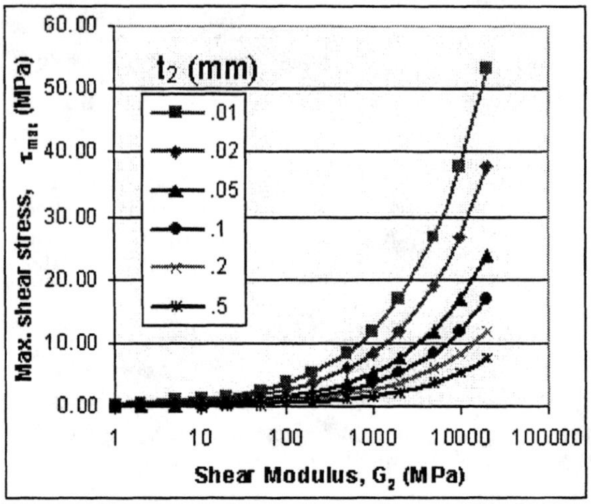

Figure 4 Maximum shear stress in the middle layer as a function of its modulus and thickness

The decrease in internal stresses due to material mismatch has led to many compliant materials (also called "low stress") being developed. The decrease in the package deformation and stresses has been quantified experimentally. For example, die-attach process experiments have been carried out using both high stress and low stress materials and analyzed using moiré interferometry [5]. The results show that the manufacturing stresses using the high stress material are on average five times higher than the case where low stress materials are used.

3.0 Design of Compliant Packages

As described above, one of the fundamental design concepts of compliant packages is the mechanical de-coupling of the die from the board, which in turn enables the use of small solder balls at fine pitch in a CSP format. This implies that there is relative movement of the die with respect to the board. Hence the design of the electrical connections between the die and the board becomes critical, as these have to withstand these movements. Hence, the design of compliant packages focuses mostly on the design of these interconnections.

3.1 CSP interconnect design

Two types of compliant packages are chosen to illustrate the design of die to substrate interconnections that can withstand the relative motion. One of them is a CSP and the other is a Wafer Level Package (WLP). The first package is a μBGA® package and is shown in

2. Leads. These interconnections are found in many package types such as Thin Small Outline Package (TSOP), Plastic Quad Flat Pack (PQFP), Plastic Leaded Chip Carrier (PLCC), Thin-Tape Carrier Package (TCP), etc. These connections exist only along the periphery of the package and the leads are flexible enough to ensure reliability.
3. Ball grid array. Plastic Ball Grid Array (PBGA), Ceramic Ball Grid array (CBGA), Flip-Chip Ball Grid Array (FC-BGA) and many other packages use this type of interconnection. These interconnections are becoming increasingly common as they provide connections over an area at a fine pitch and the package is cheaper than the pin grid array package. As the solder balls are not as flexible as the pins or leads, this package cannot depend on the flexibility of the package-to-board interconnections to provide reliability.

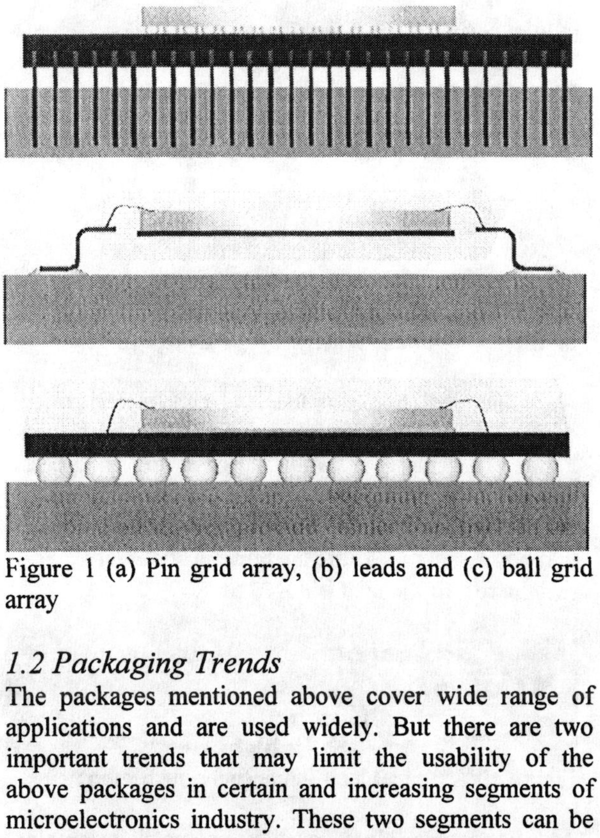

Figure 1 (a) Pin grid array, (b) leads and (c) ball grid array

1.2 Packaging Trends

The packages mentioned above cover wide range of applications and are used widely. But there are two important trends that may limit the usability of the above packages in certain and increasing segments of microelectronics industry. These two segments can be classified according to the following requirements:

1. IC die based requirements. As the die becomes more complex and the size of the features becomes smaller, the number of I/O connections increase and the distance between these connections (die pad pitch) decreases. For the package electrical performance to keep up with the die, the package must provide a short path from the die pad to the board. This usually means the package size and the package-to-board interconnections pitch must decrease.
2. Package size based requirements. With increasing trends towards miniaturization of all systems such as computers, laptops, handheld devices such as PDAs and phones, high-density servers, etc. the package size is critical. Package size is usually reduced by keeping the die to board connections such as solder balls directly under the die and also by distributing them as an array rather than just along the periphery. Hence the package size is reduced to be almost equal to the die size and these packages are called Chip Scale Packages (CSP). Ensuring reliability is more challenging in this case as neither the effect of the silicon die can be isolated by keeping the solder balls away from the die nor the solder ball size be increased due to package size and pitch restrictions. A compliant packaging concept is a very attractive solution for this class of packages.

2.0 Compliant packaging concept

Microelectronics packages are meant to provide two main functions and they are,

1. Redistribution of electrical contacts from the die pads to the solder pads on the board. As the die pad pitch is typically significantly smaller than the solder ball pitch, this redistribution is necessary and is a function of the package.
2. A structure that protects the die mechanically under different conditions encountered during its use such as temperature variation, humidity, vibration, etc. The package has to meet the electrical requirements under these conditions and this means maintaining the integrity of the silicon die, the interconnects from the silicon die pads to the substrate, the redistribution, the solder balls, etc.

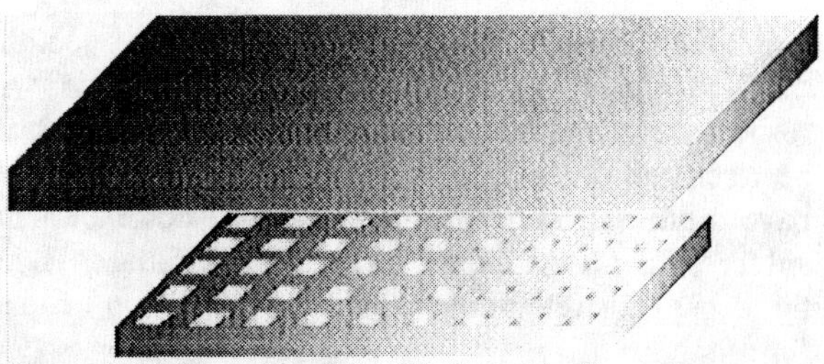

Figure 2 (a) Board and (b) IC die with die pads

Two components that are connected by the package are shown in Figure 2. As mentioned earlier, the function of the package is to electrically connect these two while satisfying mechanical reliability requirements. Packaging techniques differ in the method used to accomplish this task.

Figure 5. This package is constructed such that the active surface of the die faces the solder balls. This reduces the electrical connection length from the die pads to the solder balls, hence providing excellent electrical performance. A typical cross-section is as shown in Figure 6 and the interconnects can be either leads or wires. These two types of packages are analyzed in detail in [6,7]. The WLP is a WAVE^{TM} (Wide Area Vertical Expansion) package and it is shown in Figure 7 with the cross-section in Figure 8. Other types of compliant packages such as shown in Figure 9 [8,9] are not analyzed here.

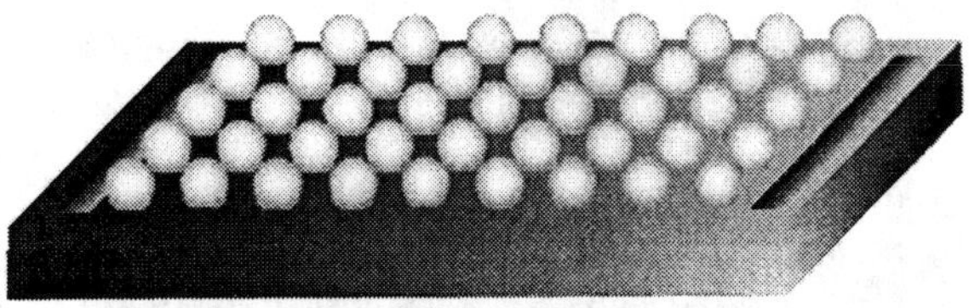

Figure 5 μBGA package with perimeter die pads

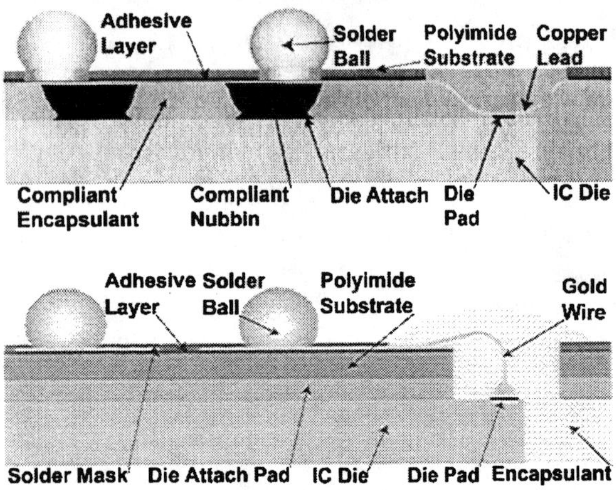

Figure 6 A schematic of a typical cross-section of a μBGA package with (a) lead bonds and (b) wire bonds

All these connections have to be designed such that they withstand the relative motion of the die with respect to the substrate due to the compliant layer between the die and the substrate. As the solder balls have minimal strain, these packages do not experience solder ball related failure but only lead failure. Hence the design for reliability reduces to the design of leads for reliability.

3.2 μBGA Lead design

For the μBGA package shown in Figures 5 and 6, the design of the interconnection shown in Figure 6(a) will be discussed in detail as the bond wire in Figure 6(b) has the required slack to withstand the relative motion.

Some of the parameters that have a significant effect on lead reliability are, (i) nubbin height, (ii) lead length and (iii) lead shape. The nubbin height requirement is governed by,

1. the compliance requirement of the package.
2. the distance required by the bonding tool to break the lead during the bonding process
3. the shaping of the lead during the bonding process

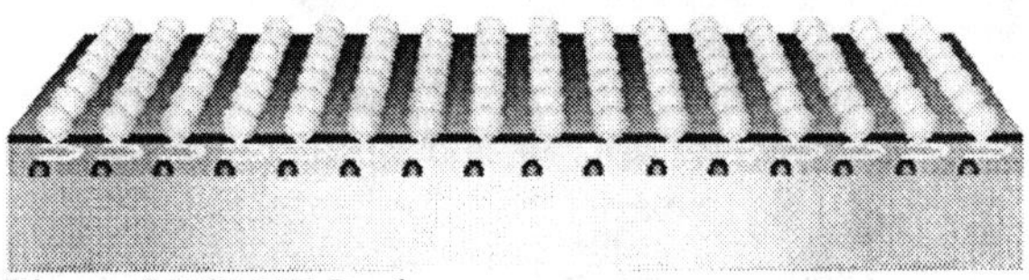

Figure 7 WAVE Package

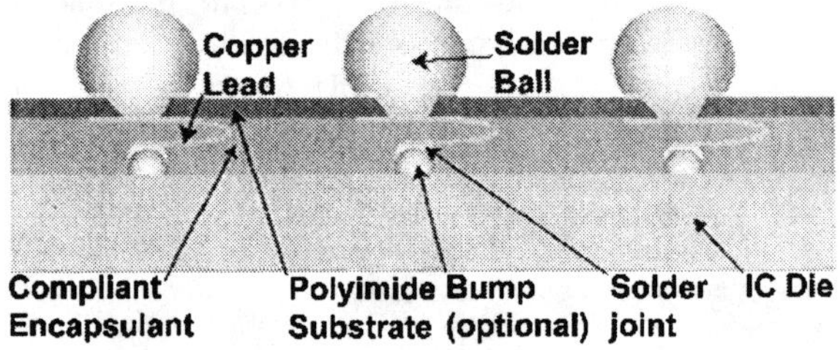

Figure 8 A cross-section of a WAVE package

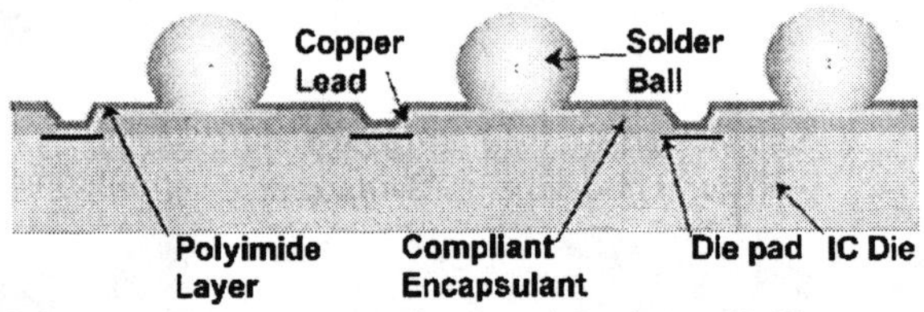

Figure 9 Compliant Wafer Level Package [8,9]

Also, the final shape of the lead can be controlled at design by designing the shape of the lead as shown in Figure 10. During bonding process, the tool presses down the lead and breaks it at the notch and bonds it to the die pad as shown in Figure 12. Typically, the notch width is less than 24 μm, anchor and shoulder is 70 μm, and the rest of the lead is 40-50 μm.

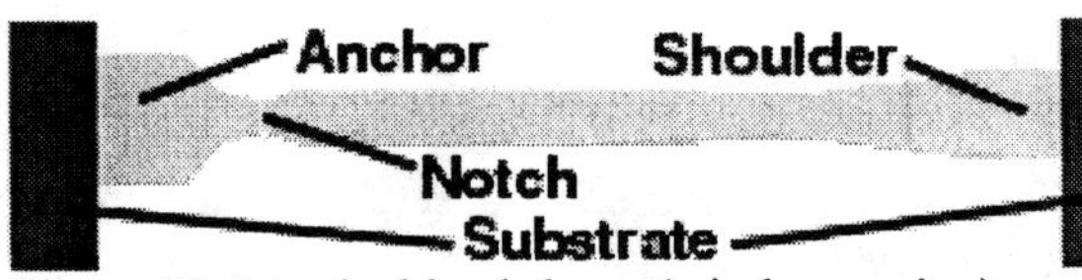

Figure 10 A typical lead shape (window region)

A design methodology based on geometric principles to determine the length, L and the height H has been developed [10]. It is based on an assumed shape as shown in Figure 11, where radii, R1 and R2 are calculated to yield an optimal shape for the given die size and the location of the leads. This theory is useful as a first approximation and more detailed analysis as given in this paper needs to be carried out for optimization.

To obtain an optimal lead shape, the lead shape (Figure 10), the tool motion (Figure 12) and the tool tip (Figure 13(a)) must be optimized. The lead shape is designed so that the curvature is low at the heel and shoulder, thus minimizing strain at these two locations. The bonding tool has a grove to "trap" the lead so that it does not slip during the lead shaping process. The shape of the tool tip is chosen to minimize the heel strain in the lead during bonding. The tool motion (the distances, CUT, TIP and Overtravel) is used to shape the lead to the desired form. The final shape of a typical bonded lead is as shown in Figure 13(b).

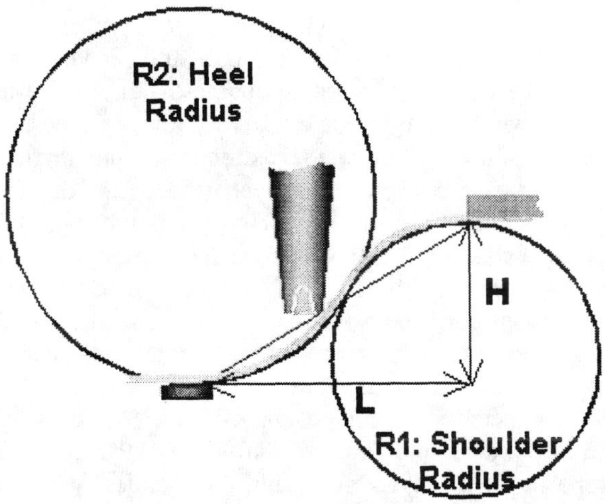

Figure 11 Lead shape based on geometric model

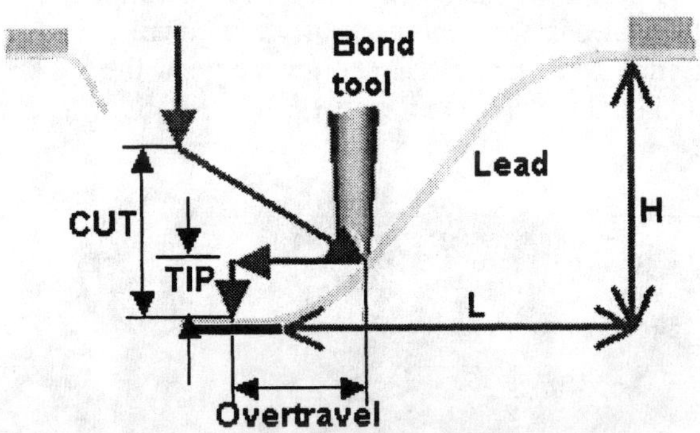

Figure 12 The lead breaking and shaping process

Figure 13 (a) The tip of the bonding tool and (b,c) SEM pictures of bonded leads

3.3 WAVE Lead design

The WAVE package shown in Figures 7-8 is a WLP and is significantly different from the μBGA package, which is a CSP. As the lead design is critical for the reliability of these compliant packages, these two types of packages have fundamentally different lead designs and processes. Detailed analysis of WAVE packages is presented in [11,12]. For WAVE package, there is no opening in the substrate as in the case of the μBGA packages. The lead lies in the plane of the substrate and during lead joining process, it is released from the substrate and bonded using solder reflow, one of many joining techniques. Hence, a bonding tool cannot be used to shape the lead that can withstand the relative motion.

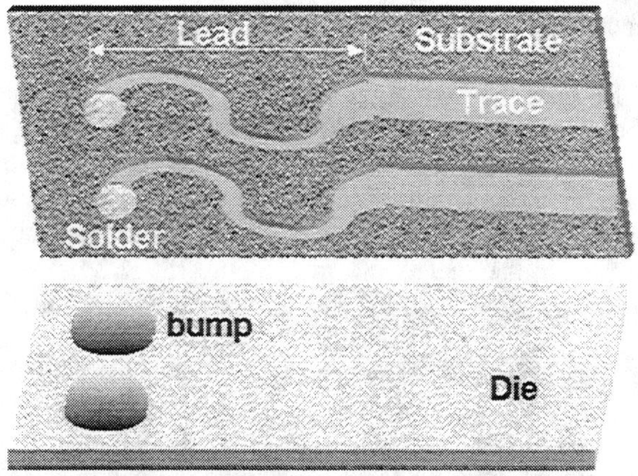

Figure 14 substrate with leads and the bumped die

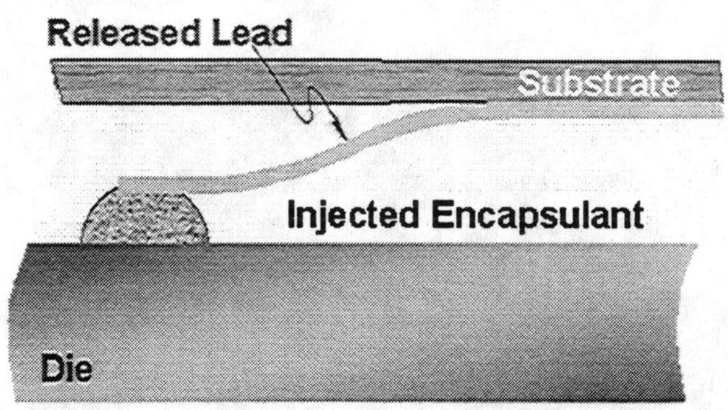

Figure 15 The shape of the lead after package assembly

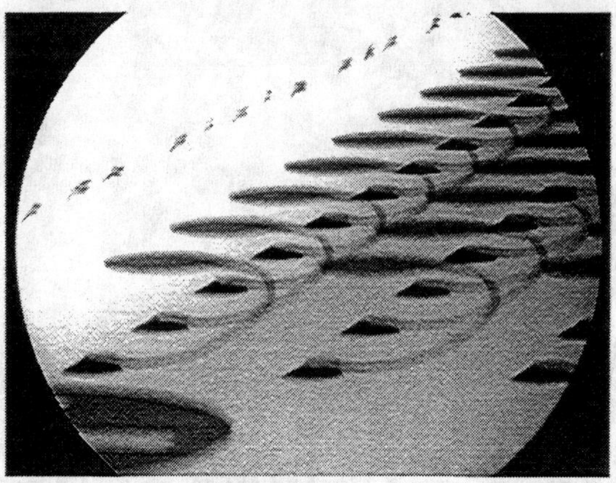

Figure 16 An X-Ray picture of the leads in WAVE package

As shown in Figure 14, the leads are essentially traces on the substrate. The tips of these leads are plated with solder, brought in contact with the bumped die, solder is reflowed to form the joint, lifted and injected with compliant encapsulant. The finished cross-section is shown in Figure 15. The leads in a finished package are shown in Figure 16.

Three lead designs for the die pad pitch of 250 μm are shown in Figure 17. One design parameter that has an effect on reliability is the amount of slack that is present in the lead. The slack is defined as the difference between the length of the lead and the straight line connecting points A and B as shown in Figure 18. Using this definition, it is seen from Figure 17 that the S-lead has the least amount of slack and the U-lead has the most amount of slack. A lead must have enough slack to withstand the lifting process and the subsequent fatigue motion during thermal cycling. The second measure of reliability is the minimum radius (R) that is present in the lead as shown in Figure 18. This radius is primarily governed by the motion in the X and Y directions, which in turn depends on the distance from the neutral point. Hence a large die would require a large radius. Lead orientation is another parameter that can be optimized for reliability [12]. These three design criteria set the limit for the minimum die pad pitch that can be handled by a WAVE package.

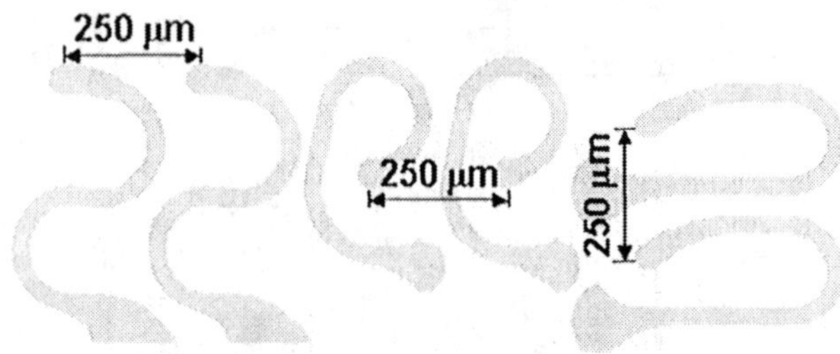

Figure 17 (a) S lead, (b) C lead and (c) U lead

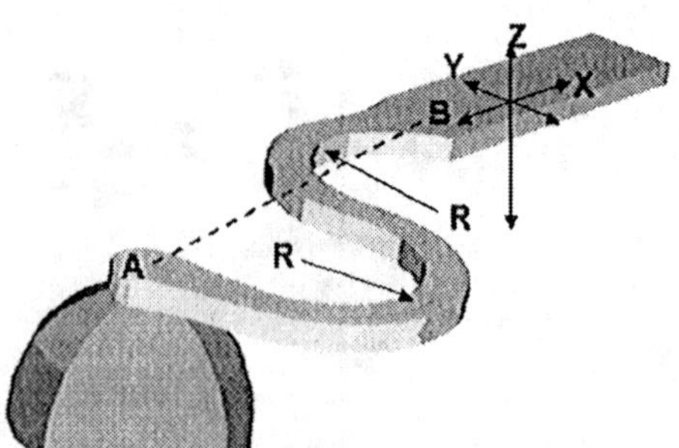

Figure 18 Lead design parameters

The leads shown in Figure 17 are best for a single or dual rows of die pads. Other leads have been designed for dense area-array with pad pitch as low as 225μm X 225μm. These leads can also be scaled for other area-array pad pitches with some minor changes to satisfy design rules of the substrate.

4.0 Reliability Analysis

The main idea behind the compliant packaging concept is to enhance reliability, especially when the package dimensions approach that of the die dimensions. In this paper, these two compliant packages were studied using finite element method as well as by carrying out reliability tests. In this section, computational reliability analysis will be presented.

4.1 μBGA Lead Analysis

A detailed analysis of μBGA packages is presented in [6]. It was shown that the solder ball strain is very low and hence it is not a potential source of failure. It was also shown that the package has very low warping and the die is nearly flat at the extremes of the thermal cycling due to the compliant layer between the die and the substrate. Due to symmetry, half the package (package size is 9.5 mm X 16.5 mm) was modeled. The package was assumed to be stress-free at the encapsulant curing temperature (150°C) and the solder balls were assumed to be stress-free at the eutectic solder solidifying temperature (183°C). The analysis was conducted by thermal cycling from 125°C to –40°C and all the results are for –40°C. All the material properties were taken from published values, with the compliant encapsulant having a modulus of 5 MPa and a CTE of 240 ppm/°C (DC6811). The solder balls and copper leads were modeled as temperature dependent elastic-plastic materials. The geometry of the package and board are given in Figure 19.

Figure 19 μBGA Package with FR4 test board

The package model was used to make sure that solder ball stress is very low and calculate the overall warping of the die and the package. As there were no individual leads in the model, the lead deformation at the two ends of the lead was determined by calculating the relative motion between points A and B shown in Figure 20.

The change in this length at -40 ^{0}C for the leads on both sides along the polyimide opening (window) when the package is on or off board are plotted in Figure 21. It is seen that the for all cases, the distance between A and B decreases and hence the lead is under compression. This indicates that the lead should be designed to handle compressive motion rather than tensile motion. Finally, the deformation near the leads was also determined so that a more refined local analysis of the leads can be carried out by using these results as the boundary conditions.

$$\Delta S = \sqrt{(L+u)^2 + v^2 + (H+v)^2} - \sqrt{L^2 + H^2} \quad (3)$$

The lead model is a unit cell model with appropriate boundary conditions, as shown in Figure 22. The breaking distance is a function of the notch dimension, the position of the notch and the tool position relative to the notch. These parameters can be varied to reduce the breaking distance. As it was found that the lead is in compression (Figure 21), overtravel is not required as the overtravel is generally done to increase the slack in the lead.

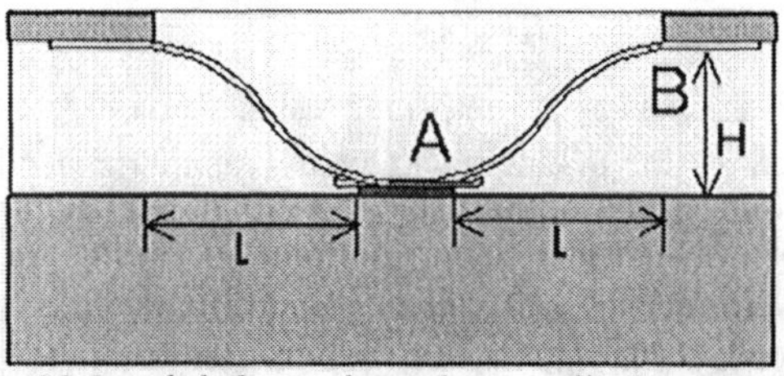

Figure 20 Lead deformation taken as displacement of B with respect to A.

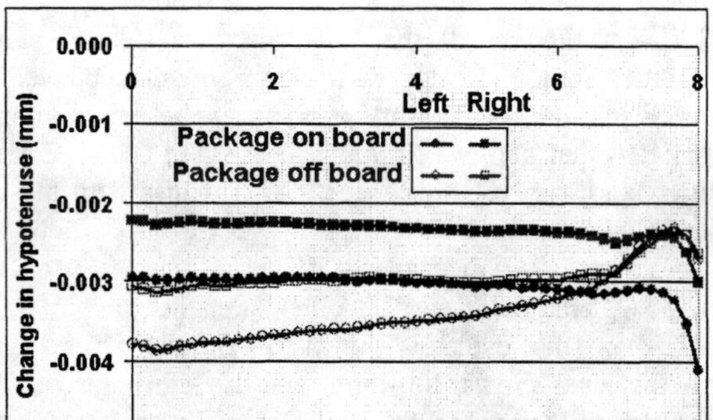

Figure 21 Deformation of the leads from equation (3)

Once the bonding process is done, the last remaining analysis is the fatigue one. In this case, the lead with the stress state as given by the bonding process is taken and the displacements obtained from the window deformation of the package models are applied cyclically to simulate thermal cycling. The lead strain states and deformed shapes are given in Figure 23. Two cases are presented in Figure 24 where the lead with DC6811 encapsulant and H = 125 μm is compared to the case with DC 6812 (E = 2.5 MPa, CTE = 195 ppm/°C) and H = 100 μm. It is clear from Figure 24 that the second case is preferable. But it was also found by analysis and experimentation that the lead is difficult to break with H = 100 μm. Even though it is better to have a lower value of H for center bond lead, the lead breaking requirement sets a lower limit on H

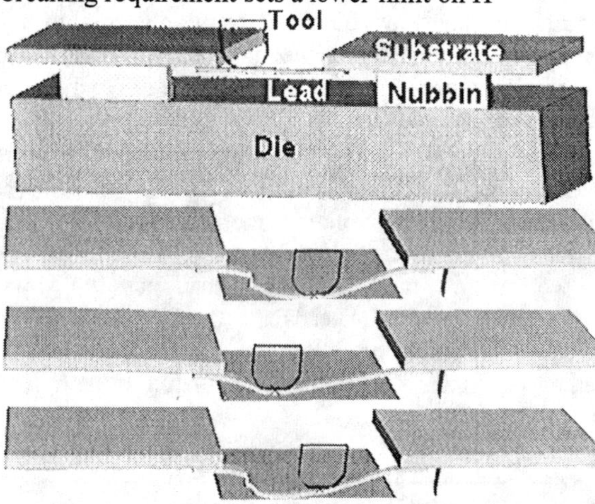

Figure 22 Lead breaking and bonding process simulation

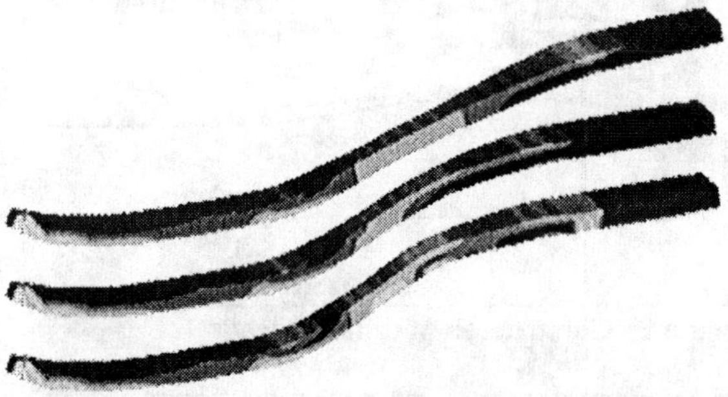
Figure 23 Lead strain state after lead bonding process, after 1 thermal cycle and after 20 thermal cycles, respectively.

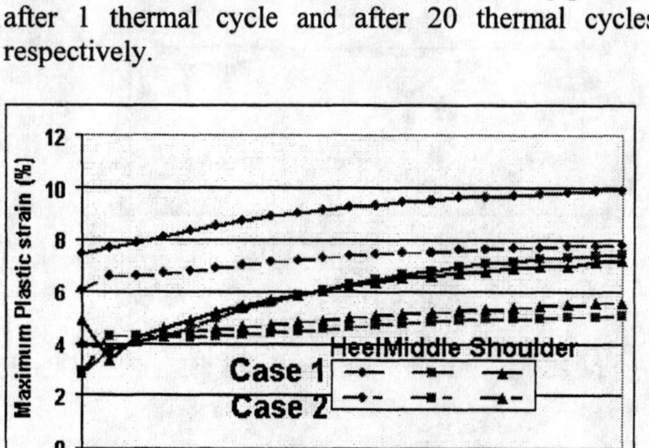

Figure 24 Lead strain state at 3 locations on lead as a function of number of thermal cycles.

4.2 WAVE Lead Analysis

A comprehensive computational analysis has been carried out and is still ongoing to characterize the leads for WAVE packages. Analysis for WAVE package was carried out similar to that of the μBGA package explained in the previous section. At the package level analysis, it was again found that the solder ball stress and the package warping are very low due to the compliant layer. After analyzing a package on board and obtaining the lead displacements, the different leads were benchmarked for reliability performance. Each lead was lifted 100 μm (encapsulant thickness) and was given an in-plane displacement of 10 μm in different directions, hence simulating the lead positioned at different points along the periphery of the die. As can be seen from Figure 25, the S-lead has the worst performance for all the leads and the U-lead is one of the best performing one. This is due to the fact that S-lead as significantly lower slack compared to the U-lead as explained in section 3.3.

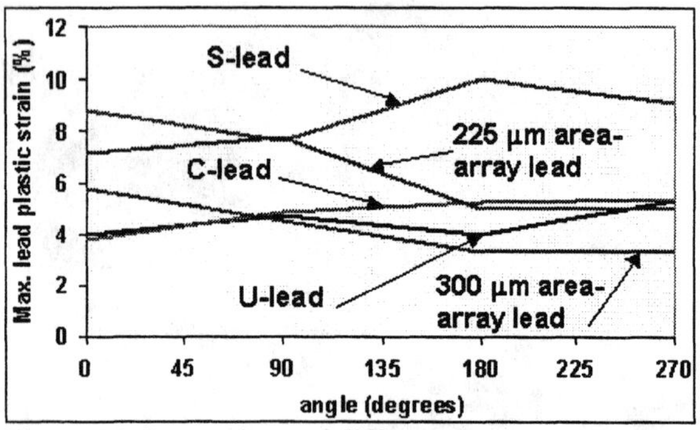

Figure 25 Comparison of different leads

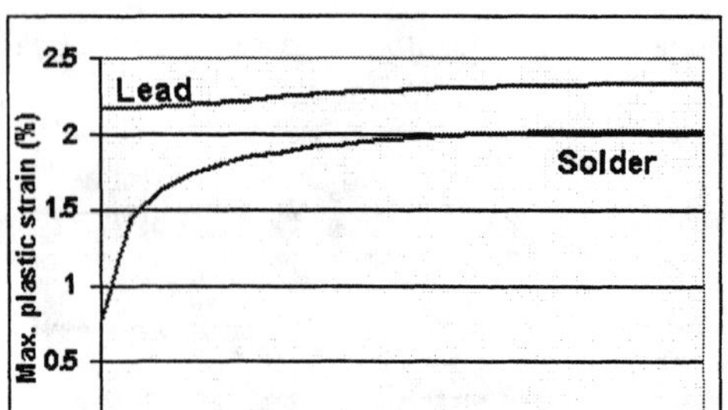

Figure 26 Maximum plastic strain in the lead and the in the solder that forms the joint.

This lead was further analyzed under fatigue loading as part of a package undergoing thermal cycling from 125 ^{0}C to –40 ^{0}C. From Figure 26, it is seen that the lead and the solder that joins the lead have a maximum plastic strain of about 2.2% and .75% after assembly process (N=0) and the strain steadily increases for the next 20 cycles. After 20 cycles, the strain ramp for the lead is 0.002%/cycle and almost 0%/cycle for the solder. Figure 27 shows the strain distribution in the lead and C4 after 20 cycles obtained through finite element analysis. It is seen that the maximum strain location is in the lead at the point of bump joint. More analysis is being planned to determine the effect of the encapsulant height, encapsulant material, bump height, other lead designs, etc.

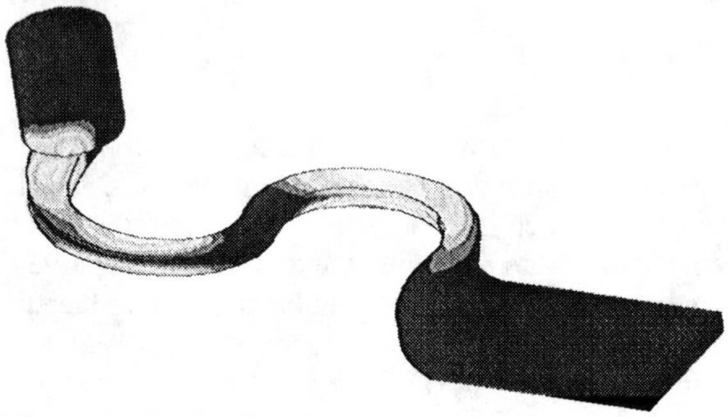

Figure 27 Strain distribution in the lead after 20 thermal cycles

5.0 Reliability Testing

The packages were built and tested for reliability. The results for the μBGA package are summarized in Table 1. This table includes many standard tests apart from the thermal cycling test that was analyzed using FEA. The package passed all normal industry requirements by a wide margin. For example, the first failure during –40°C/125°C thermal cycling occurred well over 3000 cycles, even though the internal requirement is 1000 cycles. A typical failure distribution is plotted in Figure 27. Even though all the reliability test data is not presented here, there are was good agreement between the FEA and reliability testing. The simulation predicted that the lower nubbin height and DC6812 are better and the reliability tests also confirmed these trends.

An SEM picture of a μBGA lead that has undergone more than 1000 cycles is shown in Figure 29. It is seen that this lead shows the same type of deformation and eventual failure mode as the FEA in Figure 23. This type of failure is characteristic of a lead that is in compression as was found from the package analysis results given in Figure 21.

Experiments were also carried out to determine the reliability of the WAVE packages on board that had all 3 leads shown in Figure 17. Up to 1500 cycles, only the S-lead failed as shown in Figure 29. This again agrees with the simulation results where it was predicted that the S-lead has the worst performance among all the leads. An SEM picture of a failed part is shown in Figure 31. The failure site agrees with the FEA prediction shown in Figure 27 where the maximum lead strain occurs at the bump joint. The WAVE packages are still in development stage. The samples tested here are experimental lots and hence improvement in these results is expected. The results are ongoing to determine the failure of other leads.

Table 1: Summary of reliability testing results

Reliability Test	Duration	Results (fails/total)
Package preconditioning Jedec level 1	N/A	Passed
Temperature cycle board level (-40°C ↔125°C)	1000 cycles	0/83
Temperature cycle package level (-65°C ↔150°C)	600 cycles	0/90
Temperature humidity (85°C/85% RH no bias)	1000 hours	0/66
High temperature storage life (150°C)	1000 hours	0/66
Autoclave (121°C / 100% RH, 2 atmosphere)	168 hours	0/66

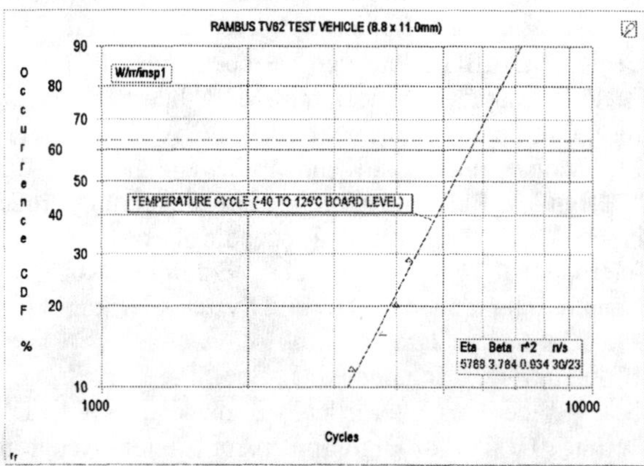

Figure 28 Wiebull plot for the current package under – 40°C ↔ 125°C thermal cycling

5.0 Conclusions

A compliant packaging concept is presented that is different from the traditional methods of making reliable microelectronic packages. From a tri-layer beam analysis, it was found that the compliance can be achieved by either reducing the modulus or by increasing the thickness of the middle layer. But for packages with thickness constraints, low modulus material is the favored option.

Figure 29 Lead kinking after more than 1000 thermal cycles before failure

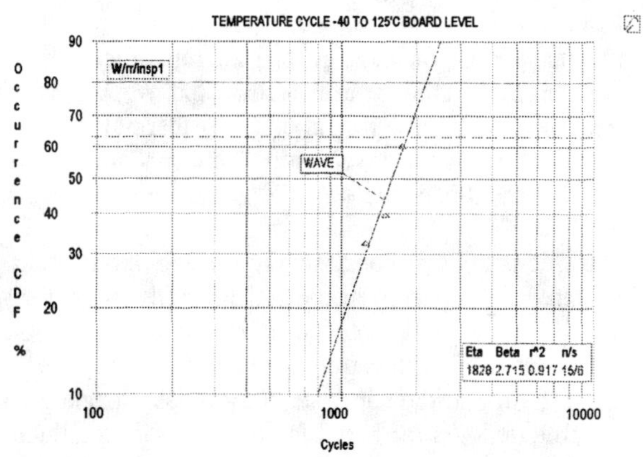

Figure 30 Wiebull plot for the S-lead

One of the features of the compliant packages is that there is relative movement between the die and the substrate and hence the connection between the die and substrate has to be able to withstand this movement. Hence, design of these interconnections (leads) determine the reliability of the compliant packages. Two types of packages, µBGA and WAVE packages were considered and their main lead design features were discussed. For the µBGA lead, the main design parameters are the nubbin height, encapsulant material, tool travel, lead length and shape. For the WAVE lead, the main design parameters are the encapsulant material, encapsulant thickness, lead slack, lead minimum radius and lead orientation.

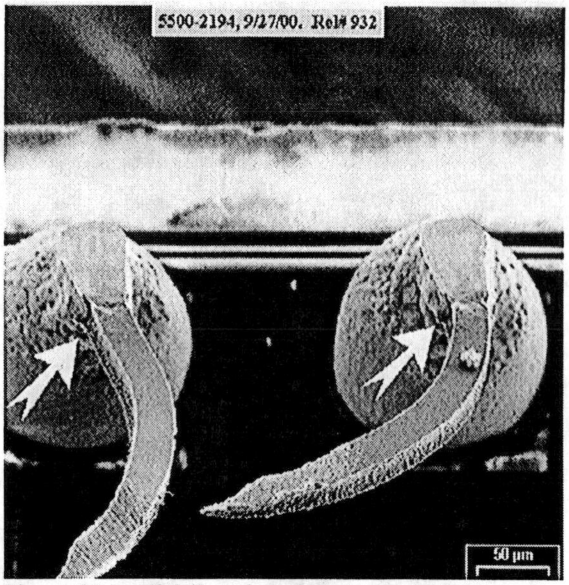

Figure 31 Lead failure location

Reliability analysis using FEA was conducted for both types of packages. Solder ball strain and the package warping were found to be minimal. The reliability testing and failure analysis confirmed the analysis results. μBGA package passed all reliability tests by a wide margin. One type of lead in the WAVE package experienced failure just after 1000 thermal cycles. As these packages are still under development, improvement is expected.

Thus, it is shown that designing of compliant packages requires a shift from traditional methods of analysis as the package behavior and failure modes are totally different. But with careful design of interconnections from the die to the substrate, very reliable packages can be obtained. As the trend in packaging is to reduce the size of the package to almost that of the silicon die, using a low modulus encapsulant to make a compliant package is a very attractive method of ensuring package reliability.

6.0 References

1. Lau, John, "Ball Grid Array technology", McGraw-Hill, Inc. 1995.
2. Timoshenko, S. P. and Goodier, J. N., "Theory of Elasticity", McGraw-Hill, Inc.
3. Suhir, E., "Interfacial Stresses in Bimaterial Thermostats", ASME Journal of Applied Mechanics, vol. 56, pp. 595-599, 1989.
4. Hokanson, K.E. and Bar-Cohen, A., "A Shear Based Optimization of Adhesive Thickness for Die Bonding", IEEE Transactions on Components, Packaging and Manufacturing Technology, Part A, Vol. 18, No. 3, pp. 578-584, 1995.
5. Tsao, P.-H. and Voloshin, A.S., "Manufacturing Stresses in the Die due to the Die-Attach Process", IEEE Transactions on Components, Packaging and Manufacturing Technology, Part A, Vol. 18, No. 1, pp. 201-205, 1995.
6. Mohammed, I. and Kim, Y-G, "Design and Reliability Study of Center Bonded Lead μBGA Package for RAMBUS DRAM", Proceedings of the Advanced Technology Workshop for Chip Scale Packages, North Falmouth, MA, September 22-23, 2000.
7. Mohammed, I., Sridhar, K. and Kim, Y-G, "Design and Reliability Study of Wire-Bonded μBGA Packages for High Performance DRAM Applications", to be published in the 2001 Proceedings of 34th International Microelectronics And Packaging Society (IMAPS) Symposium, Baltimore, MD, Oct. 9-11, 2001.
8. Kazama, A., Satoh, T., Yamaguchi, Y., Anjoh, I. And Nishimura, A., "Development of Low-Cost and highly Reliable Wafer Process Package", 2001 Proceedings of the 51st Electrical Component and Technology Conference (ECTC), pp. 40-46, Orlando, FL, 2001.
9. Bakir, M.S., Patel, C.S., Kohl, P.A., Martin, K.P. and Meindl, J.D., "Ultra High I/O Density Package: Sea of Leads (SOL)", Proceedings of the 2001 International Conference and Exhibition of High Density Interconnect (HDI) and Systems Packaging, pp. 335-339, Santa Clara, CA, 2001.
10. Young Kim, David Gibson, and Mike Warner, "CSP Design and Reliability: Lead Design Guide for Center Pad μBGA Package", Proceedings of The Chip Scale Packaging Symposium, SMTA International, San Jose, CA, Sept. 12-16, 1999, p. 89-96.
11. Mohammed, I., Kim, Y-G. and Kang, T-G., "Design and Reliability Study of Lead Bonding Process in WAVE Packages", Proceedings of the 2001 International Conference and Exhibition of High Density Interconnect (HDI) and Systems Packaging, pp. 233-238, Santa Clara, CA, 2001.
12 Kim, Y-G., Mohammed, I., Seol, B-S. and Kang, T-G., "Wide Area Vertical Expansion (WAVE™) Package Design for High Speed Application: Reliability and Performance", 2001 Proceedings of the 51st Electrical Component and Technology Conference (ECTC), pp. 54-61, Orlando, FL, 2001.

Proceedings of
2001 ASME International Mechanical Engineering Congress and Exposition
November 11–16, 2001, New York, NY
EPP-Vol. 1

IMECE2001/EPP-24706

COMPLIANT WAFER LEVEL PACKAGE

Chirag Patel
IBM
Yorktown Heights, NY

Kevin P. Martin
Georgia Inst. Of Technology
Atlanta, GA

James D. Meindl
Georgia Inst. Of Technology
Atlanta, GA

ABSTRACT

A high I/O density and high performance wafer level packaging technology called the Compliant Wafer Level Package (CWLP) is reported. The necessity for compliant interconnects in upcoming generations of electronic products is discussed by analyzing the technology requirements projected by the International Technology Roadmap for Semiconductors (ITRS). To be a true wafer level package, the technology should have following three characteristics[11]: I) package all Integrated Circuits (ICs) intact on wafer at once, II) perform wafer level test and burn-in, and III) assemble the WLP on the system board without using an underfill. Compliant interconnects are essential to accomplishing wafer level test and assembly without underfill. These topics are discussed in the paper followed by fabrication and performance analysis of the CWLP technology.

INTRODUCTION

Chip Scale Packages (CSP) have been recognized as an enabling technology for future miniaturized and highly functional electronic products. The wafer level packages (WLPs) are an obvious extension in the evolution of the CSP technology, yet the WLPs have the capability of revolutionizing and in many ways restructuring the conventional package, assembly, and testing processes used in the electronic industry. Wafer level package (WLP), often described as a package-less package because of its chip size dimensions, offers several key benefits in terms of packing efficiency, electrical performance, and cost (package and testing). Many different types of wafer level package technologies have been reported in the literature[1-10]. These WLPs are mostly fabricated in two ways: one employs additional substrates such as polymeric tape, silicon or glass material[1,6] and the others use thin film sequential processing on top of the wafer[2-5,10]. Based on the method of their fabrication, the WLP technologies reported thus far can be classified into serial and parallel approaches. The serial approach simply adds additional photolithography steps or redistribution layers to the wafer to package it. The redistribution layer is used to re-route the chip I/Os to a larger pitch and then the solder bumps are fabricated at the wafer level by using the standard C4 technology. In the parallel approach, the wafer level package is fabricated independently of the wafer and the WLP is attached to the wafer by a wafer level joining methodology[6]. The mixture of these two approaches can also be found where substrate materials are employed in conjunction with sequential thin film processing[1].

To be a true wafer level package, the technology should have following three characteristics[11]: I) package all Integrated Circuits (ICs) intact on wafer at once, II) perform wafer level test and burn-in, and III) assemble the WLP on the system board without using an underfill. Compliant interconnects are essential to accomplishing wafer level test and assembly without underfill. The first of the three characteristics, i.e. packaging all of the ICs intact on wafer at once, is available in all of the WLP technologies reported in the literature. However, the majority of the WLPs do not provide the ability to conduct wafer level test and burn-in or allow system assembly without an underfill for many applications. Without addressing these two characteristics, the WLP technology will slightly modify the conventional packaging process and it will have negligible impact on electronics packaging. Other shortcomings of the WLP technology at present are its low input/output (I/O) count, heat removal inefficiency, poor reliability, and difficulty in integration with the semiconductor and the system board manufacturers.

HIGH I/O COUNT AND HIGH PERFORMANCE TECHNOLOGY REQUIREMENTS

Many of the chip packaging requirements (e.g. technical performance, physical specifications, and manufacturing costs) are spelled out in the International Technology Roadmap for Semiconductors (ITRS)[12] prepared by the semiconductor industry. The ITRS projects packaging requirements for six different categories: low cost, hand-held, cost-performance, high-performance, automotive and memory. The most demanding requirements for bandwidth, I/O count, and cost are captured in Figure 1 through Figure 3, respectively.

For high performance market, the off-chip speed or bandwidth is expected to approach giga-hertz while the I/O count required for highly functional products is projected to increase monotonically beyond five thousand per chip. The cost per I/O is projected to be as low as 0.30 cents per pin. The memory and low cost applications outline the lowest cost requirements while the automotive sector presents severest heat removal specifications. The hand-held and portable electronics require highest packing efficiency (i.e. size, height and weight of the package). In addition to these specific IC needs, the package must be able to easily integrate with semiconductor and system board manufacturers without requiring massive re-designs.

CWLP FABRICATION

The CWLP fabrication process consists of three photolithographic masking steps. The fabrication process starts at the completion of IC fabrication when the die pads are exposed on the wafer. In the CWLP process, there is no boundary between the IC and the package; in contrast, the package fabrication can be described as a continuous process of the IC fabrication. The prototype described in this section was primarily designed to establish and optimize CWLP process flow and characterize the electrical and mechanical performance of the package. Based on the results of this experiment, more advanced versions of CWLP in terms of high I/O density were developed and are presented later in the paper. In this prototype, 3-inch silicon wafers were used as substrates. The silicon wafer was fabricated with simple test structures as shown in Figure 4. The test devices were aluminum die pads, 75 µm by 75 µm, connected with 10 µm wide metal interconnect. Each die consisted of 512 die pads and 256 metal lines connecting them.

The pads were placed at a pitch of 440 µm over 1 cm^2 die area. The prototype wafer contained total of 32 die. The wafer was sputter coated with 0.7 µm thick *Al* layer and etched to reflect the die pads and interconnect patterns as drawn in Figure 4. The test patterns had two objectives: (a) mimic the IC die pads to mark the completion of IC fabrication, and (b) enable electrical characterization and functional testing of the package as described later in the paper. The CWLP fabrication process is described hereafter. The process is schematically shown and the steps involved in the fabrication process are described in Figure 5. Following the completion of the test structures, the wafer was coated with 25 µm thick layer of a low modulus polymeric material. The polymer was chosen for its low stress properties, Young's Elastic Modulus of 0.89 GPa, and low moisture absorbance, less than 0.5%. The polymer was a mixture of polyimide and siloxane, where 35% of the polymeric solvent contained siloxane. Using photolithographic mask, the polymer was patterned and carefully etched to expose the die pads and to yield a sloped sidewall of approximately 75° – 80°. Standard polymer curing in nitrogen purged convective furnace followed this step. After the polymer cure, the wafer was sputter coated with Titanium/Copper (*Ti/Cu*), 18-nm/500 nm, metal seed layer to be used for electroplating copper compliant interconnects. Using second photolithographic mask, copper was electroplated on the seed layer to form the compliant interconnects (also referred to as leads). Copper electroplating bath was made from Cupric Sulfate ($CuSO_4$) and Sulfuric Acid (H_2SO_4). The bath was kept at room temperature and the solution was constantly stirred at 2000 rpm. The final thickness of the electroplated compliant interconnect was 10 µm. The compliant interconnect width was 50 µm. In this one step, all of the compliant leads were fabricated monolithically, i.e. all of the I/Os used for electrical and mechanical connection between the chip and the system board were fabricated at once. The seed layer used for compliant interconnects was removed and the wafer was again coated with 15 µm of low modulus polymer layer to embed the compliant leads. Using third and the last photolithographic mask and aforementioned etching procedures, the polymer was patterned and cured to expose the end of the interconnect. The wafer was sputter coated with 18 nm/ 500 nm, *Ti/Cu*, seed layer to be used for electroplating eutectic *Sn/Pb* solder bump. Eutectic *Sn/Pb* was electroplated to the thickness of 35 µm. The *Sn/Pb* bath was kept at 50°C during the plating process and was constantly stirred at 2000 rpm. The removal of the *Ti/Cu* seed layer used for electroplating the solder bumps marked the completion of the CWLP fabrication process. With only three photolithographic masks, all of the ICs intact in wafer form were fully packaged and all of the compliant interconnects were monolithically fabricated in one step.

NECESSITY FOR COMPLIANCE

Wafer level packages are expected to meet the following key criteria: I) package all ICs intact on wafer at once, II) possess the ability to perform wafer level test and burn-in, and III) ability to be assembled on the system board without the use of an underfill. An omission of any of the criterion will revert the newly developed technology back to already existing versions of CSP and Flip Chip technologies.

By using standard photolithographic fabrication process, the CWLP approach is able to package all of the ICs intact on wafer at once. The polymer used in the CWLP, also referred to as interposer in the paper, had three main functions: (a) to passivate and encapsulate the ICs on wafer from atmosphere and UV particles, (b) to provide low stress medium such that the compliant interconnect can flex under thermo-mechanical stress, and (c) to be compliant so that it can easily be compressed when a force is applied. The flexible interconnects were a crucial part of the package and were specifically designed to be compliant in horizontal, *xy*, and vertical, *z*, directions to enable system assembly without using underfill and wafer level test and burn-in. An underfill is used to absorb the thermo-mechanical stress arising from the coefficient of thermal expansion (CTE) mismatch between the system board and the silicon chip. Underfills require long processing time and when applied it is difficult to rework the assembled part. These two factors make it a costly solution. In the case of the CWLP, the compliant interconnects physically stretch and contract as required during temperature cycling to absorb the thermo-mechanical stress. In this way, monolithically fabricated compliant leads not only provide electro-mechanical connection from the chip to the system board, but they also replace the costly underfill process.

Furthermore, the leads and the interposer were also designed to be compliant in vertical direction. Vertical compliance was necessary to be able to burn-in and perform reliability and AC/DC functional testing at wafer level. Because of non-planarity of test probe cards, it is required to press the card against the wafer to make contact to all of the I/Os. If there were no vertical compliance, as in the case of flip chip devices, the contact is made to the silicon die pads and can jeopardize the functionality of the sensitive silicon devices. In case of the CWLP, the contact is being made to the eutectic solder bump, 50 μm above the silicon surface.

The compliant interconnect physically deflect in vertical direction as needed to make firm contact to all of the I/Os on the wafer. The combination of three dimensional compliance and fully packaged wafer allowed wafer level burn-in, reliability and functional testing. At the end of these tests at wafer level, the wafer was diced into individual chips and assembled on the system board without underfill.

CWLP TEST DEVICES AND MECHANICAL COMPLIANCE CHARACTERIZATION

In the experiment, many different types of compliant leads were fabricated to test for their electrical and mechanical properties. Figure 6 shows one of the fabricated test patterns where the die pads, signal trace for electrical testing, polymer via, the compliant interconnect, and the solder bump attachment point are labeled in the figure. The compliant interconnects with serpentine shape in Figure 6 were mechanically affixed at the die pad. When assembled on the system board, the PWB attachment pad extends and contracts to absorb thermal stress as discussed in the previous section. To measure this physical movement, an apparatus was set up to make contact to solder bump at PWB attachment pad and apply predetermined mechanical force. The lead was held at its contracted or extracted position and the distance from its original position was recorded as lateral compliance of the lead. A cycle comprised of expansion-original-contraction-original position of the lead. The lead was measured to expand and contract ± 50 μm and always returned to its original position without any failure for 1500 cycles. Simple calculation was made to assess the total compliance required to address the CTE mismatch between FR-4 board (17 ppm/°C) and silicon chip (3 ppm/°C). The difference of 14 ppm/°C was multiplied by 100 °C to give maximum difference in expansion of 14 μm per 1 cm package. The ITRS projects largest chip to be 2.75 cm on side in year 2010. The CWLP is chip size package and therefore the compliant leads must expand at least (14 × 2.75 = 38.5 μm). Although more testing must be done, the preliminary results on the mechanical compliance of CWLP lead indicate 50 μm of expansion for 1500 cycles.

More variations of the compliant interconnects were also fabricated on the same prototype wafer to determine the best design in terms of yielding the most mechanical compliance. These patterns are shown in Figure 7. The failures during reliability testing were more frequently recorded in the leads that had multiple arcs. This was attributed to the fact that the mechanical stresses were mostly accumulated near the PWB attachment pad in the lead during experimental simulations. This portion of the lead contributed to *x, y* movement. In the case of multiple arcs (Figure 7(e)), the pad was observed to collide with the legs of the arc during mechanical movement. The arcs were essential as they evenly distributed the stress along the lead; however, the optimum number has to be determined by theoretical and further practical simulations. The compliant leads that have sharp corners (Figure 7(f)) also performed poorly mainly because the corners became a point for maximum stress accumulation.

Final CWLP dimensions for the first prototype were exactly same as the chip 1 cm by 1 cm, making it a true wafer level package. The package height excluding the wafer was measured to be 60 μm. The eutectic *Sn/Pb* bumps were re-flowed during system assembly and the package was bonded to FR-4 printed wiring board. The package mounting area on the board was also same as the package area. All of the 512 I/Os in 1 cm^2 area were routed within single layer of the FR-4 board using 100 μm lines and spaces. The connections were routed to the periphery of the mounting area on the board and then extended further out to enable electrical testing before and after system reliability testing. The board layout is shown in Figure 8. The CWLP has 100% silicon to package efficiency and similarly requires the same area on the PWB, reducing the size of the PWB and in turn, the final product itself.

HIGH I/O DENSITY CWLP

As the functionality of the electronic products increases, higher number of input/output (I/O) connections will be required. To avoid the problems with reliability and usage of underfill during system assembly, most of the existing wafer level package technologies have been applied to low I/O count devices such as memory and integrated passive components. The CWLP can easily be scaled to extremely high I/O count with its area array distribution and minimum geometry compliant lead designs. Many different types of compliant leads shown in

Figure 7 were characterized for their electrical and mechanical properties before and after the reliability testing. The leads that showed best performance without failure through the reliability testing was optimized and used to build the next high density I/O CWLP. The first prototype only had 512 I/Os as it was mainly designed for test and characterization purposes. As a next step to this prototype, the CWLP with 1000 I/Os per 1 cm^2 area was developed using the same fabrication methodology described in Figure 5. The only design change was the reduction of I/O pitch to 310 μm in an area array. The 1000 I/O CWLP is carefully designed to provide the same characteristics (i.e. three dimensional compliance, system assembly without underfill, good reliability, excellent electrical and mechanical performance) as low I/O count CWLP described earlier. The SEM images of high density CWLP are shown in Figure 9 (a,b). In these images, the second layer of polymer coating is omitted to get a better understanding of the structure. Figure 9(a) shows a close up SEM image of the compliant leads and the array of 1000 leads per 1 cm^2 is shown in Figure 9(b). The ITRS projections[12] indicate that packages with density of 1000 I/Os per 1 cm^2 will be required for high performance ICs by year 2014. However, at present such high I/O density can be used for improving electrical clock and power distribution on chip and also for electrical test.

CONCLUSIONS

The necessity of compliant interconnects is essential in reducing the complexity and cost of testing. The compliant interconnects are also a viable alternative to non-underfill assembly. Ease of fabrication (at wafer level) and high reliability are crucial factors in making the compliant interconnects an integral part of the electronics packaging. A compliant wafer level package (CWLP) technology was reported that can: I) package all ICs intact on wafer at once, II) possess the ability to perform wafer level test and burn-in, and III) ability to be assembled on the system board without the use of an underfill. Mechanical characterizations of CWLP prototype were presented to show its high reliability design. The package was shown to have high packing efficiency and high I/O density. A CWLP with 1000 I/Os per 1 cm^2 area was also developed and reported that used the same fabrication process as low I/O CWLP. Furthermore, the high I/O density CWLP was carefully designed to provide the same characteristics (i.e. three dimensional compliance, system assembly without underfill, good reliability, excellent electrical and mechanical performance) as low I/O count CWLP.

REFERENCES

1. A. Badihi and E. Por, "ShellCASE- A True Miniature Integrated Circuit Package," *Proc. Intl. FC, BGA, AP Symposium*, California, February 1995, pp. 244-252.
2. J. Young and H. Clearfield, "Wafer-Level Packaging for Low-Cost, Low-Lead-Count Devices," *2nd Annual Semiconductor Packaging Symposium*, SEMICON West, July 1999.
3. H. Eslampour and T. Hamano, "Wafer Level Packaging Development and Super CSP Status," *Chip Scale International*, September 1999.
4. I. S. Park, et al., "Wafer-Level CSP (Omega-CSP)," *Chip Scale International*, September 1999.
5. J. Lau, "A Novel and Reliable Wafer-Level Chi Scale Package (WLCSP)," *Chip Scale International*, September 1999.
6. J. Fjelstad, "W.A.V.E. technology for wafer level packaging of ICs," *Proc. of the Electronic Packaging Technology Conference (EPTC)*, pp. 214-218, Singapore, December 8-10, 1998.
7. B. Haba, M. Beroz, and J. W. Smith, "Wafer Level Packaging: Limitations And Solutions," *Chip Scale International*, San Jose, September 1999.
8. J. Healy, "Microspring: The Ineluctable Inductive Construct for High Performance Wafer Scale Packaging," *International Packaging Strategy Symposium*, SEMICON West, July 15 1999.
9. C. S. Patel, et al., "Low Cost High Density Compliant Wafer Level Package," *International Conference on High-Density Interconnect and Systems Packaging,* Denver, Colorado, April 26-28, pp. 262-268, 2000.
10. S. Savastiouk, O. Siniaguine, E. Korczynski, "3D Wafer Level Packaging," *International Conference on High-Density Interconnect and Systems Packaging,* Denver, Colorado, April 26-28, pp. 26-31, 2000.
11. Philip E. Garrou, "Status and Prospects of Wafer Level Chip Scale Packaging", *Plenary Session, 49th Electronic Components and Technology Conference (ECTC)*, June 1999.
12. "International Technology Road Map for Semiconductors (ITRS) 1998 update", http://notes.sematech.org/NTRS/Rdmpmem.nsf/pages/MHome.htm

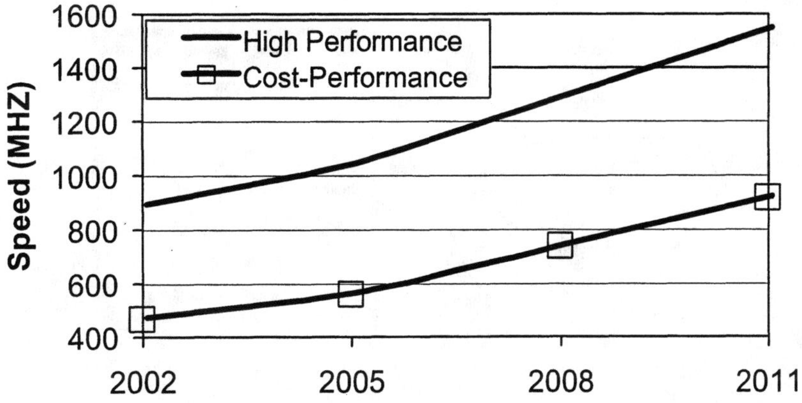

Figure 1: Chip-to-chip communication speed requirement

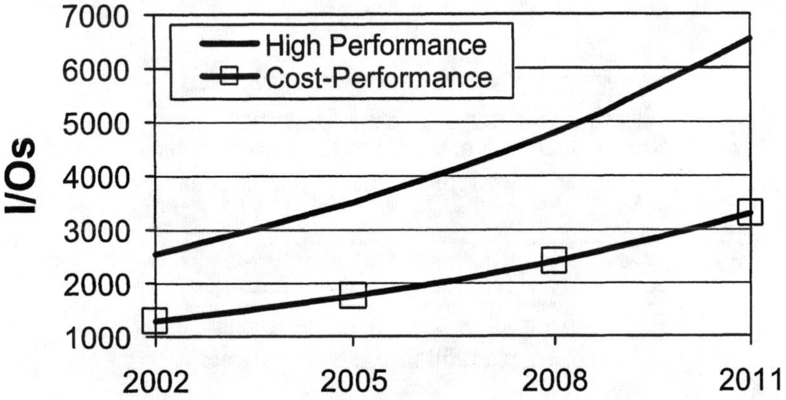

Figure 2: Maximum number of I/O Count

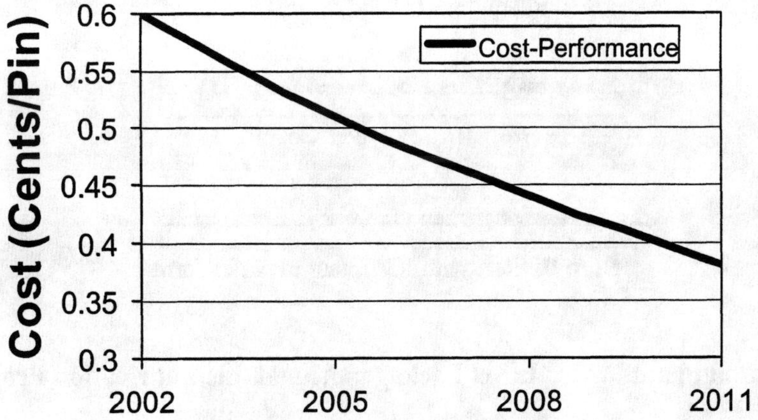

Figure 3: Maximum Cost per I/O

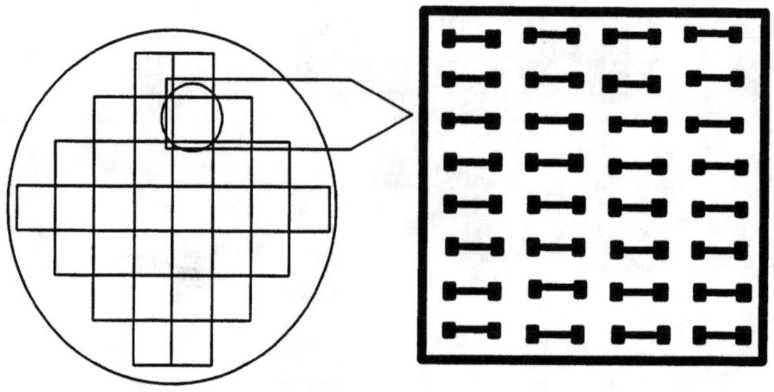

Figure 4: CWLP test prototype

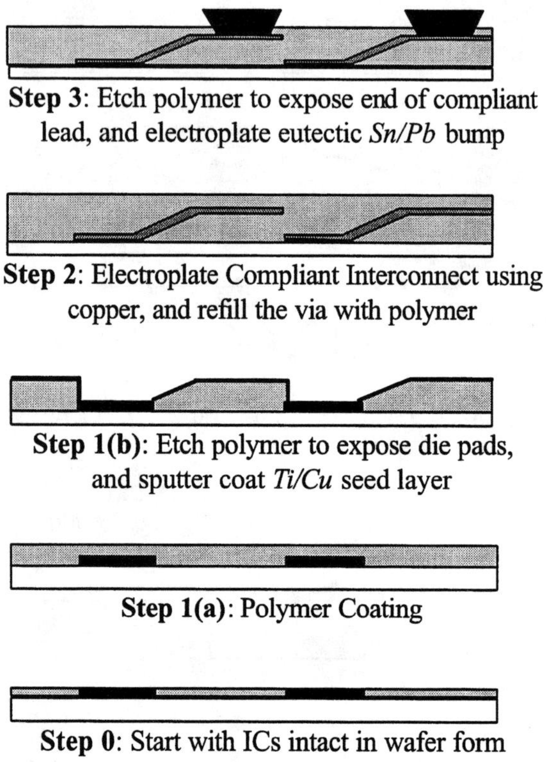

Figure 5: Compliant Wafer Level Package: Monolithic Fabrication Process Flow

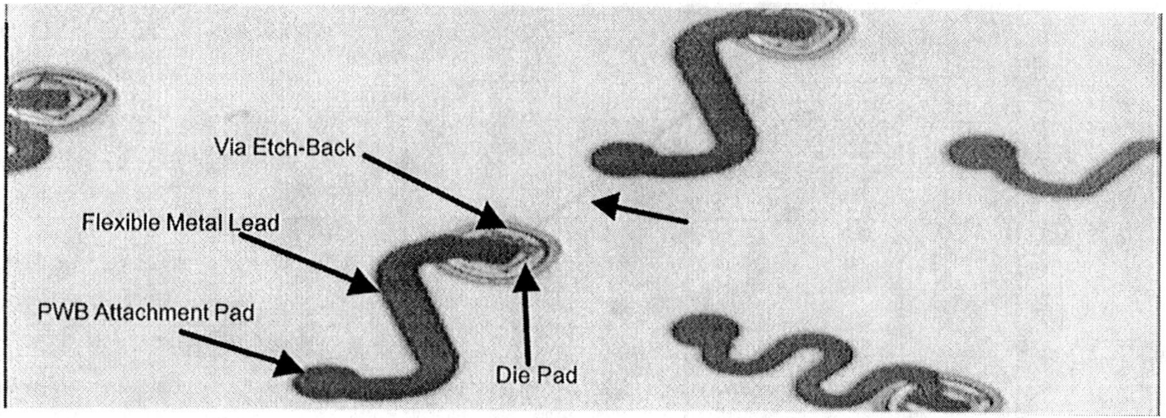

Figure 6: CWLP Prototype Test Devices

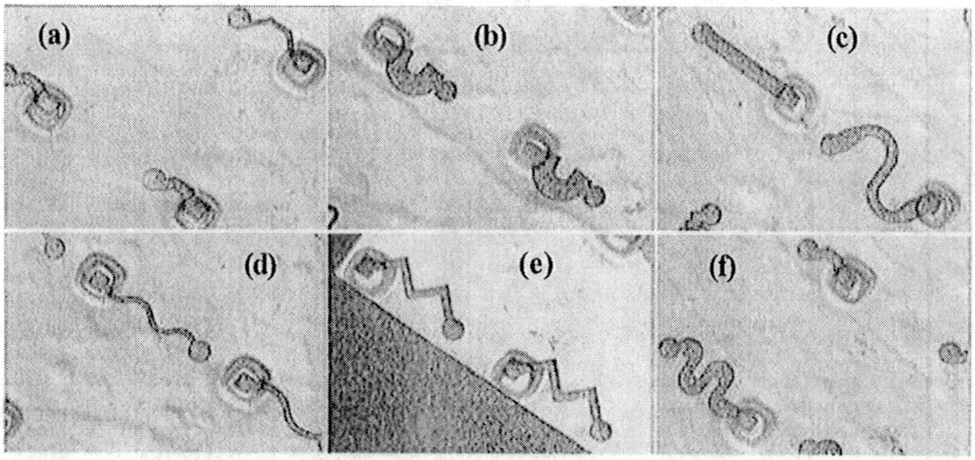

Figure 7 (a-f) : CWLP device test patterns

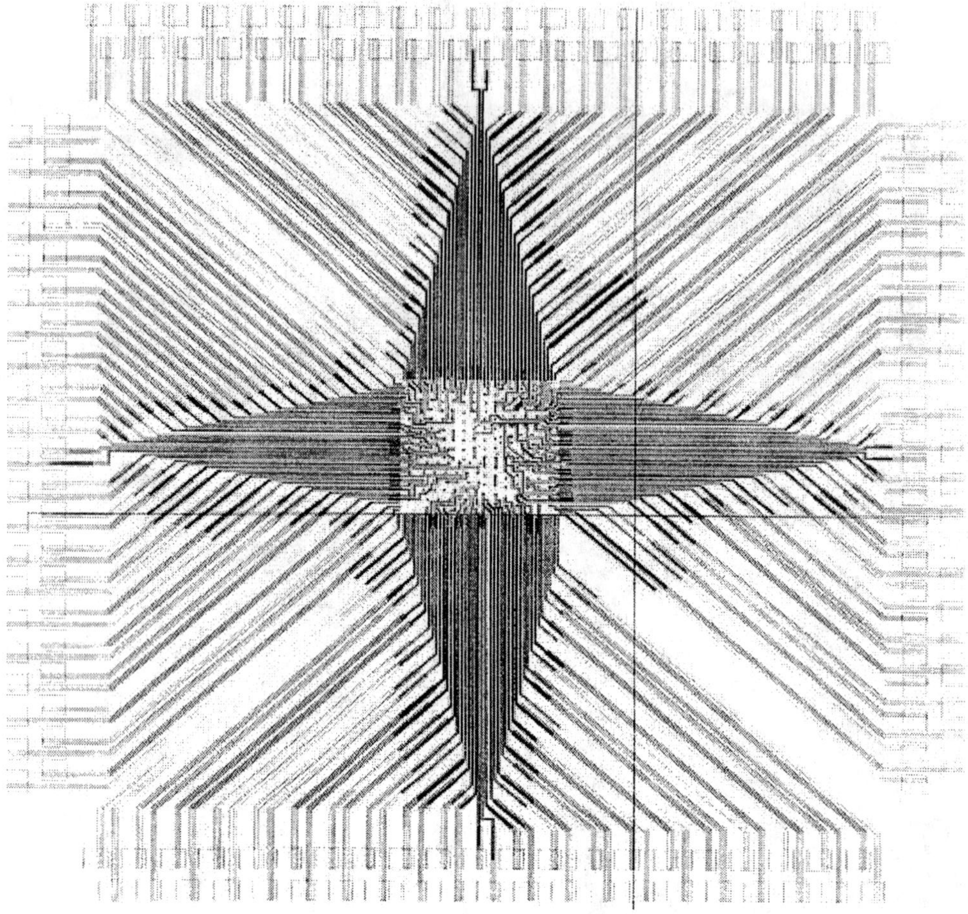

Figure 8: PWB layout for the CWLP test prototype

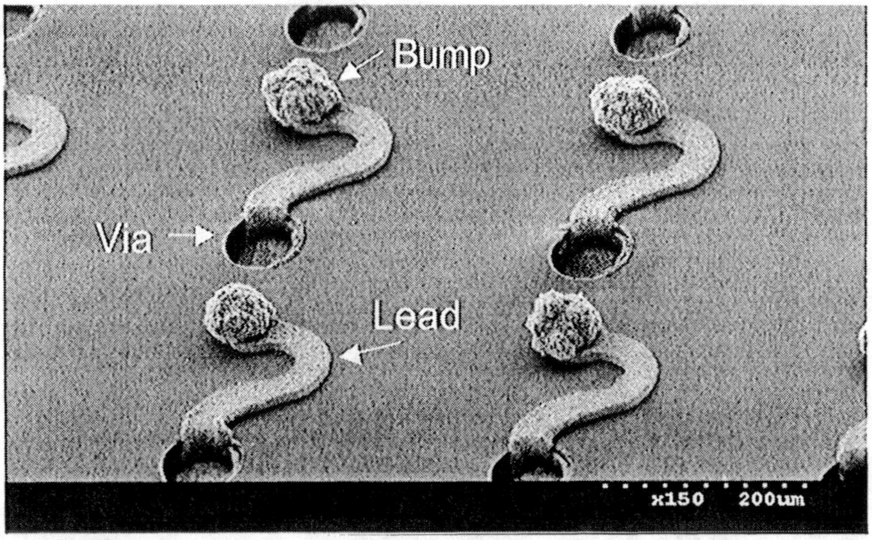

(a)

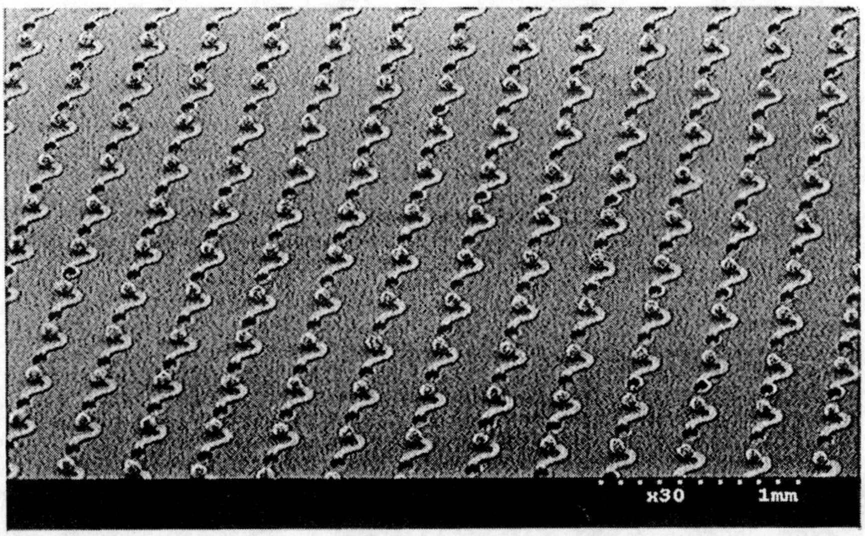

(b)

Figure 9: High Density CWLP (a) Close-up of leads, (b) area array distribution of 1000 compliant leads per 1 cm²

Proceedings of
2001 ASME International Mechanical Engineering Congress and Exposition
November 11–16, 2001, New York, NY
EPP-Vol. 1

IMECE2001/EPP-24707

NANO-MICRO SCALE CHARACTERIZATION OF DIELECTRIC MATERIALS

Paul A. Kohl and Sue Ann Bidstrup Allen
Georgia Institute of Technology
School of Chemical Engineering
Atlanta, Georgia 30332-0100
404-894-2893, paul.kohl@che.gatech.edu

ABSTRACT

Thin films of dielectric materials are used as insulating layers in microelectronic systems, including integrated circuits and printed circuit boards. Polymer and spin-on-glasses are widely used as coatings and interlevel dielectrics. The dielectric permitivity and loss are the electrical properties of most interest. However, characterization and optimization of the mechanical properties is essential to the successful implementation and usage of these insulators. In this presentation, thin-film structures for measuring direction-dependent properties (dielectric constant, elastic modulus, and coefficient of thermal expansion), and nano-indentation results on spin-on-glass materials will be presented. Significant improvement in the fracture toughness of nanoporous methylsilsesquioxane glass films has been demonstrated upon introduction of porosity into the films.

INTRODUCTION

Thin films used as interlevel dielectrics and passivation layers are essential components of advanced electronic packages [1,2]. Polymer films are typically formed by solvent casting (e.g. spin-coating, spraying or extrusion coating) or through lamination. During solvent casting, stresses are generated in the film as the solvent evaporates and the film collapses in the *through*-plane (z) direction while being constrained by the substrate in the *in*-plane (x) direction. Stresses also arise due to a mismatch in the coefficient of thermal expansion (CTE) between the constrained film and the substrate [3-6]. These residual stresses can cause polymer chain orientations [7-9] which results in different thermo-mechanical properties in the plane and normal to the plane of the film. Since thin film mechanical properties are critical for electrical and mechanical design and modeling [10,11], it is desirable to have the properties characterized *in-situ* in order to accurately determine the conditions during use. This paper will report on the in-situ measurement of through-plane properties for polymer thin-films via the dual capacitor method, and in the improvement in mechanical properties of thin glass films by means of nano-indentation.

Background

The dual capacitor technique involves the use of two devices, shown in Figure 1, as capacitive sensors to detect changes in dielectric constant and thickness of a polymer film under normal, compressive stress. The polymer was compressed by applying a *through*-plane load, and the capacitance of the devices measured as a function of load. Models were then used to calculate the thickness and dielectric permittivity of the polymer film from the measured capacitance. A response of a parallel plate capacitor (PPC) was used to measure the *through*-plane dielectric permittivity of the polymer. The response of the electrode (IDE) structures was modeled using an electromagnetic field simulator. The through-plane modulus was evaluated by applying a force to the PPC and IDE. The load experiments were conducted by compressing the polymer-coated glass substrates using an Instron and measuring the capacitance of the devices at different static loads. The measured stress-capacitance data was then used to calculate the *through*-plane strain and the changes in the *in*-plane and *through*-plane permittivities using the device models. The through-plane coefficient of thermal expansion was evaluated by thermally cycling the devices and measuring the capacitive response of the two electrodes.

Nano-indentation was used to probe the mechanical properties of porous, methylsilsesquioxane films (MSQ) spin-on-glass films. Commercially available MSQ was made porous by the use of porogen, as the sacrificial material.

RESULTS AND DISCUSSION

Commercially available polymers were used in this study. The direction-dependant dielectric properties for several materials were measured and are shown in Figure 2. This shows a plot of the in-plane (x and y direction) permittivity as a function of the through-plane (z) permittivity. Some of the polymers had identical properties in the x and y (in-plane) directions. Thus, there is only one value of permittivity reported on the abscissa of Fig. 2. Those polymers straddling the 45 degree line on Fig. 2 have isotropic behavior for in-plane (ε_x or ε_y) and through-plane properties (ε_z), $\varepsilon_x = \varepsilon_y = \varepsilon_z$. Points which lie off the 45 degree line, exhibit in-plane vs. through-plane anisotropy with ε_x not equal to ε_z. Three of the spin-coated materials in Fig. 3 (i.e. Probimide 293, PI-2611, and PI-2540) have in-plane permittivities greater their through-plane permittivity. This results from orientation of the polymer during processing and curing.

The in-plane permittivities (x and y direction) were not the same for orthotropic materials like Vectran, liquid crystal polymer (LCP) films. Two in-plane permittivity values (ε_x and ε_y) are plotted in Fig. 2 for the reinforced and non-reinforced Vectran. The nonreinforced and glass-reinforced films show a large anisotropy with a higher permittivity for the through-plane compared to the in-plane directions. The x and y directions for the reinforced and nonreinforced Vectran were identified with respect to the direction that the films were extruded. The degree of anisotropy for the glass-reinforced film in the y-direction (ε'_y) is lower than in the x-direction (ε'_x) whereas for the nonreinforced case, the results show the opposite trend. The anisotropy in the polymer chains is due to crystallization of the LCP along the direction of draw during processing. Since the modulus along the chains is much greater than the modulus normal to it, the LCP films have exceptional mechanical properties along the direction of draw but moderate otherwise.

Through-plane Modulus

IDE and PPC devices were load- tested with Cyclotene 4026 as the polymer dielectric. Figure 3 shows a typical compressive stress vs. through-plane strain. As a *through*-plane compressive stress is applied to the sample, the bow of the substrate is reduced. This causes an increase in the *in*-plane tensile stress resulting in a relatively large compressive deformation of the polymer film in the *through*-plane direction. Compressive strain is also a direct consequence of the applied stress. The dimensional and dielectric changes in the film during this part of the experiment are relatively non-linear and highly dependent upon the initial residual stress-state in the film. After the bow is mitigated, the *through*-plane strain and dielectric changes in the film occur only due to the applied stress. The deformations and volume changes during this stage are much smaller, linear, and independent of the initial state of stress in the film.

The capacitance vs. stress using a PPC was recorded for each polymer. For Cyclotene 4026, the average value of $E_{z,eff}$ was 3.17 ± 0.53 GPa. Using the measured values for $E_{z,eff}$ and Poisson's ratio equal to 0.34, the isotropic modulus E was then calculated to be 3.06 ± 0.38 GPa. This value compares well with the *in*-plane modulus of Cyclotene 4026 measured by Im *et al.* [12] using a tensile testing method as 2.9 ± 0.2 GPa and suggests that the polymer may be mechanically isotropic. This is reasonable since Cyclotene 4026 is a thermosetting glassy polymer, which has been shown to be electrically isotropic. The mechanical and electrical properties are dependent on film orientation; hence, one would expect that if isotropic behavior is found in one property it would also be found in the other.

Several PPC and IDE devices were fabricated and tested to determine the *through*-plane modulus of PI-2611. The *in*-plane and *through*-plane dielectric permittivity at different stress levels were calculated. The average effective modulus $E_{z,eff}$ measured using the PPC devices was 2.11 ± 0.69 GPa. The *through*-plane elastic modulus can be estimated to be 2.45 GPa. When compared to the *through*-plane modulus of BPDA/PDA reported in literature, the estimated modulus value of 2.45 GPa was found to be greater than the 0.05-1.1 GPa reported from the air-capacitance gauge measurements [13, 14] but less than the 6.1 GPa estimated using the hydrostatic strain measurement technique [15]. Unlike those in this report, all reported literature measurements were made on free-standing films.

Since the analysis of the stress-strain measurements from the air-capacitance gauge method disregarded the effect of volume change in the polymer film, their calculations may have underestimated the *through*-plane modulus. The modulus was calculated from the slope of the stress-strain curve and was reported to increase with applied stress. In contrast to this method, others used the change in free volume of the polymer and decreasing mobility of its chains as an explanation for their observation of increasing modulus [14]. If thickness deformation of the polymer film due to volume change were accounted for, a higher value for modulus would have be obtained.

Nano-indentation of porous Spin-on-glass

Methylsilsesquioxane (MSQ) spin-on-glass films were prepared with a porogen:MSQ ratio of 0:100, 10:90, 20:80, 30:70, and 40:60. The results show that the pore size is similar (3-10 nm), and independent of the loading level. Increasing the amount of sacrificial polymer in the sample does increase the pore density (number of pores per volume) in the film, but does not affect the pore size. The dielectric constant showed a reduction from $\varepsilon_r = 2.68$ for a non-porous MSQ film, to $\varepsilon_r = 2.24$ for a porous film of 30:70 wt.% porogen:MSQ (after decomposition).

The mechanical properties of porous materials are critical in the integration capability of these films. Nanoindentation experiments were performed to characterize the mechanical properties. Figure 4 shows an array of indents made to study the resistance to crack propagation of the films. The experiments were performed at 10, 2, 1, 0.5, 0.1, and 0.01 mN of force. Figure 4a shows indents at 10, 2, 1, and 0.5 mN force (left to right). The 0.1 and 0.01 mN indents were totally elastic and therefore, are not visible in Fig. 4a (no plastic deformation or cracking occurred). Five repetitions of each indent are shown in Fig. 4a from top to bottom. It is important to note that each indent was made in a single load-unload segment. That is, the indenter penetrated the surface once with the specific applied force, and then immediately unloaded. Figure 4b corresponds to a film of 90:10 wt.% MSQ with TESNB as the sacrificial polymer. The array of indents in this image was made exactly the same way as in the MSQ sample, with the only difference that only four repetitions (top-to-bottom) are shown. In the last row (5th from the top), every indent was made at 10 mN force. The dramatic contrast between the cracks in Fig. 4a and Fig. 4b depicts the remarkable improvement in the mechanical properties of the films with the introduction of porosity. In the case of the MSQ films, Fig. 4a, the cracks extended all the way to the edge of the wafer. With the porous films, the cracks did not propagate beyond the point of indentation. From Figure 4b it is clear that some plastic deformation occurred in both films because the impressions were retained after the indentation process. Pure MSQ is known to be susceptible to cracking due to the intrinsic stress in the film. The difference between MSQ and fully-relaxed silicon dioxide lies with the bond-strain in MSQ, where each silicon atom is bonded to 1.5 oxygen atoms, and one non-bridging CH_3 group. The effect of the intrinsic strain in these bonds imposes a thickness limit on these films. It is difficult to fabricate MSQ films thicker than about 600 nm because they crack upon processing. Although the films in Fig. 4 were below the cracking threshold for MSQ (films of ~ 400 nm), the indentation process induced cracking. The crack propagation relieved the intrinsic stress in the films.

A quantitative measure of the fracture toughness was made and is presented in Figure 5 for samples with and without porosity.. The average fracture toughness as a function of the initial concentration of sacrificial TESNB is shown in Fig. 5. The values were the average of 30 indentations in each sample at the applied forces mentioned before. The results are also shown as a function of the average thickness of the films.

The results show a 10x increase in the fracture toughness with a 10 wt.% increase in the concentration of sacrificial polymer in the film. Crack propagation is terminated at the boundary of a defect, or in the case of porous MSQ, at the boundary of a pore. With higher porosity, the films are more relaxed and there are more boundaries for crack termination. Also, the elastic modulus of the porous films is lower, thus reducing the driving force for crack initiation.

CONCLUSIONS

The dual capacitor technique was shown to be a novel way to measure *in-situ*, the *through*-plane modulus of thin polymer films. The technique utilizes two devices, an interdigitated electrode and a parallel plate capacitor, as sensors to track the change in dielectric permittivity, density and thickness of a polymer film due to an applied load. These measurements were used in a stress model, developed to predict the elastic stress-strain interactions in the polymer-substrate system, to calculate the *through*-plane modulus of the film. The derived stress model accounted for both, the change in thickness and the change in volume in the film. Results were reported for 10-14 □m films of Cyclotene 4026 and polyimide PI-2611. The *through*-plane modulus calculated for Cyclotene 4026 was 3.06 ± 0.38 GPa which when compared with its reported *in*-plane modulus of 2.9 GPa indicates an isotropic film. The technique was also used to measure the *through*-plane modulus of a highly anisotropic polymer, PI-2611. An anisotropic stress model was used to calculate the *through*-plane modulus as 2.45 ± 0.64 with the assumptions, $E_x = 9.8$ GPa, $\nu_{xy} = 0.3$ and $\nu_{xz} = 0.23$. This estimated value was greater than the *through*-plane modulus measured using the air-capacitance gauge method but lower than the value measured using a hydrostatic strain approach and an impulsive stimulated thermal scattering technique, reported in the literature. Finally, a sensitivity analysis was performed with respect to the elastic constants and it was shown that the Poisson's ratio had the largest impact on the calculated *through*-plane modulus. The uncertainty in the E_z due to the uncertainty in the input parameters, however, was less than the standard error in the experimental measurements.

For porous MSQ spin-on-glass films, high degree of porosity can be achieved by bonding the sacrificial material to the spin-on glass matrix. Films with porogen loadings of up to 40 wt.% have been fabricated without agglomeration or phase separation of the sacrificial polymer. Significant improvement in the fracture toughness of the films, and a slight decrease in the elastic modulus and the hardness were observed with increasing porosity. The index of refraction and the dielectric constant also decrease with porosity.

ACKNOWLEDGEMENTS

The contributions from Kaushal Patel, Agnes Padovani, Thomas Hodge are gratefully acknowledged. The financial support of the National Oak Ridge National Laboratory is acknowledged.

REFERENCES

1. R. R. Tummala and E. J. Ryamaszewski, Ed., *Microelectronics Packaging Handbook*, Van Nostrand-Reinhold, New York, 1997.
2. P. Garrou, *Proc. of the IEEE*, **80**, 12, 1942-1954 (1992).
3. H. M. Tong, C.-K. Hu, C. Feger, P. S. Ho, *Polym. Eng. Sci.* **26**, 17, 1213-1217 (1986).
4. C. L. Bauer, R. J. Farris, in *Polyimides: Materials, Chemistry and Characterization*, C. Feger, M. M. Khojasteh, and J. E. McGrath, Ed., Elsevier, Amsterdam, 1989, p. 549-562.
5. M. T. Pottiger, J. Coburn, *Mat. Res. Soc. Symp. Proc.*, **227**, 187-194 (1991).
6. J. C. Coburn, M. T. Pottiger, S. C. Noe, S. D. Senturia, *J. Polym. Sci. Polym. Phy.*, **32**, 1271-1283 (1994).
7. L. Lin, Ph.D. thesis, Georgia Institute of Technology, Atlanta, GA, 1994.
8. S. E. Molis, in *Polyimides: Materials, Chemistry and Characterization*, C. Feger, M. M. Khojasteh, J. E. McGrath, Ed., Elsevier, Amsterdam, 1989, p. 659-672.
9. D. Boese, H. Lee, D. Y. Yoon, J. D. Swalen, J. F. Rabolt, *J. Polym. Sci. Polym. Phy.*, **30**, 1321-1327 (1992).
10. S. Weinberg, Masters thesis, Georgia Institute of Technology, Atlanta, GA, January 1996.
11. E. T. Ryan, T. Cho, I. Malik, Z.-H. Zhao, J. K. Lee, P. S. Ho, *Mat. Res. Soc. Symp. Proc.*, **476**, 135-146 (1997).
12. J. Im, E.O. Shaffer, R. Peters, T. Rey, C. Murlick, R.L. Sammler, In *Proc. of International Conference on Microelectronics*, Reston, VA, p. 168-175, 1996.
13. T.Y. Wu, D.L. Questad, *Mat. Res.. Soc. Symp. Proc.*, **264**, 143-160 (1992).
14. H.-C. Liou, P.S. Ho, A. McKerrow, *J. Polym. Sci. B*, **36**, 1383-1392 (1998).
15. K.C. Sheth, M.J. Chen, R.J. Farris, *Mat. Res. Soc. Symp. Proc.*, **356**, 529-534 (1995).
16. B. R. Lawn, A. G. Evans, and D. B. Marshall, *J. Am. Ceram. Soc.*, **63**, 574 (1980).
17. D. S. Harding, W. C. Oliver, and G. M. Pharr, *Mat. Res. Soc. Symp. Proc.*, **356**, 663 (1995)

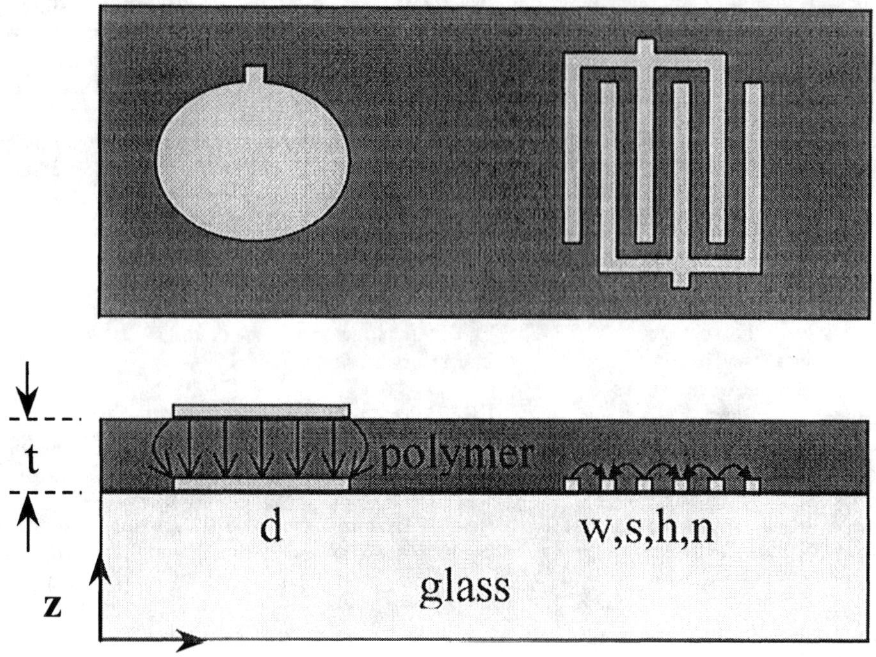

Figure 1: Dual capacitor device configuration for the anisotropic dielectric characterization of deposited polymer thin films

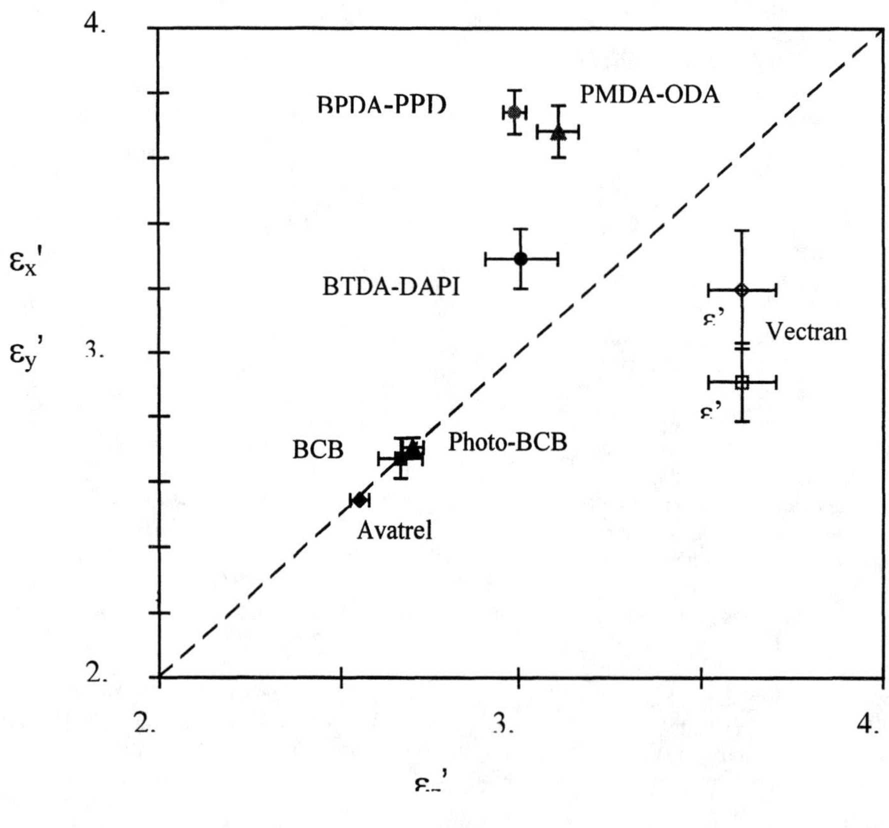

Figure 2. In plane dielectric constant (y-axis) vs. through plane (x-axis) for polymers.

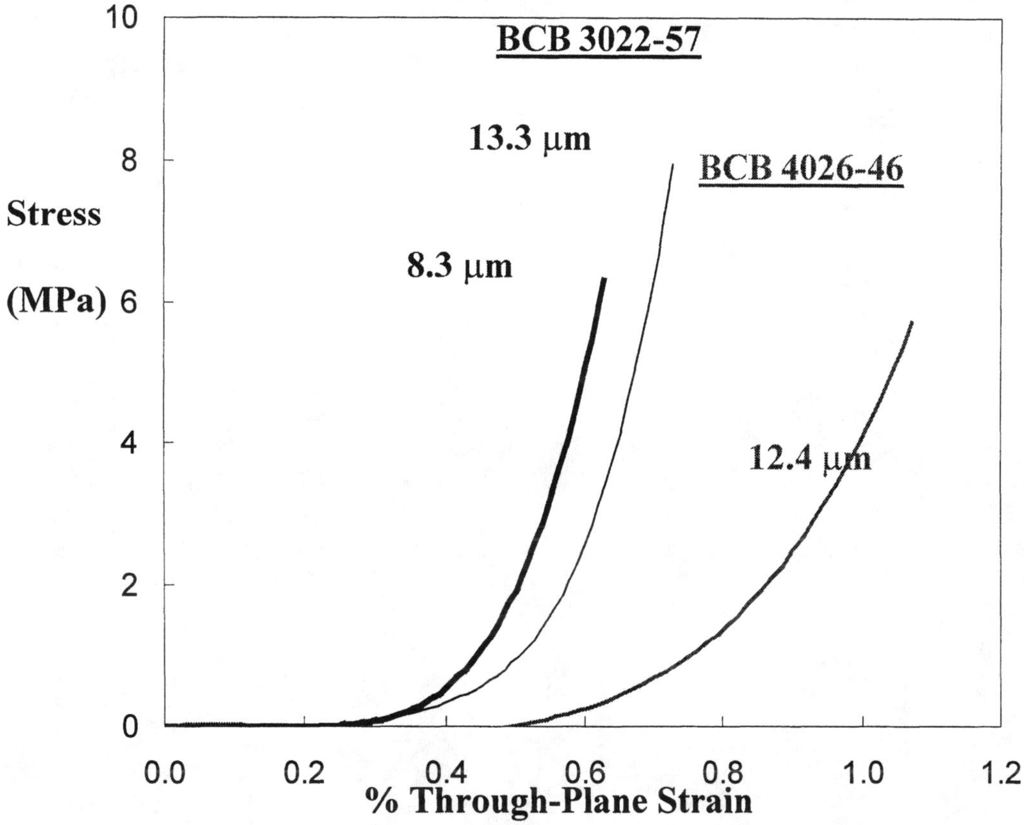

Figure 3. Stress vs strain for Cyclotene 3022 and 4026

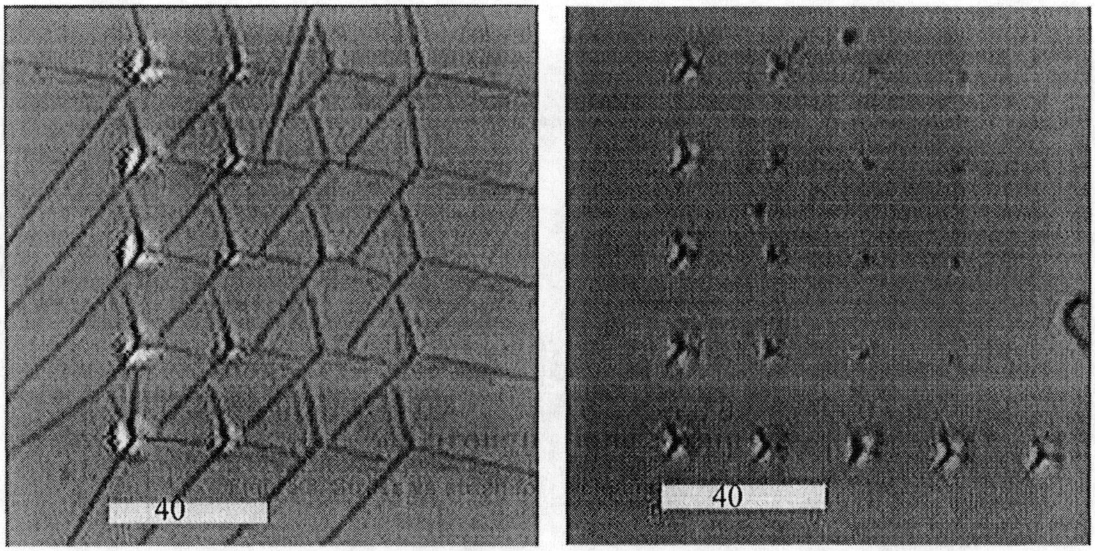

Figure 4. Optical picture of nano-indents in (a) non-porous MSQ, and (b) porous MSQ

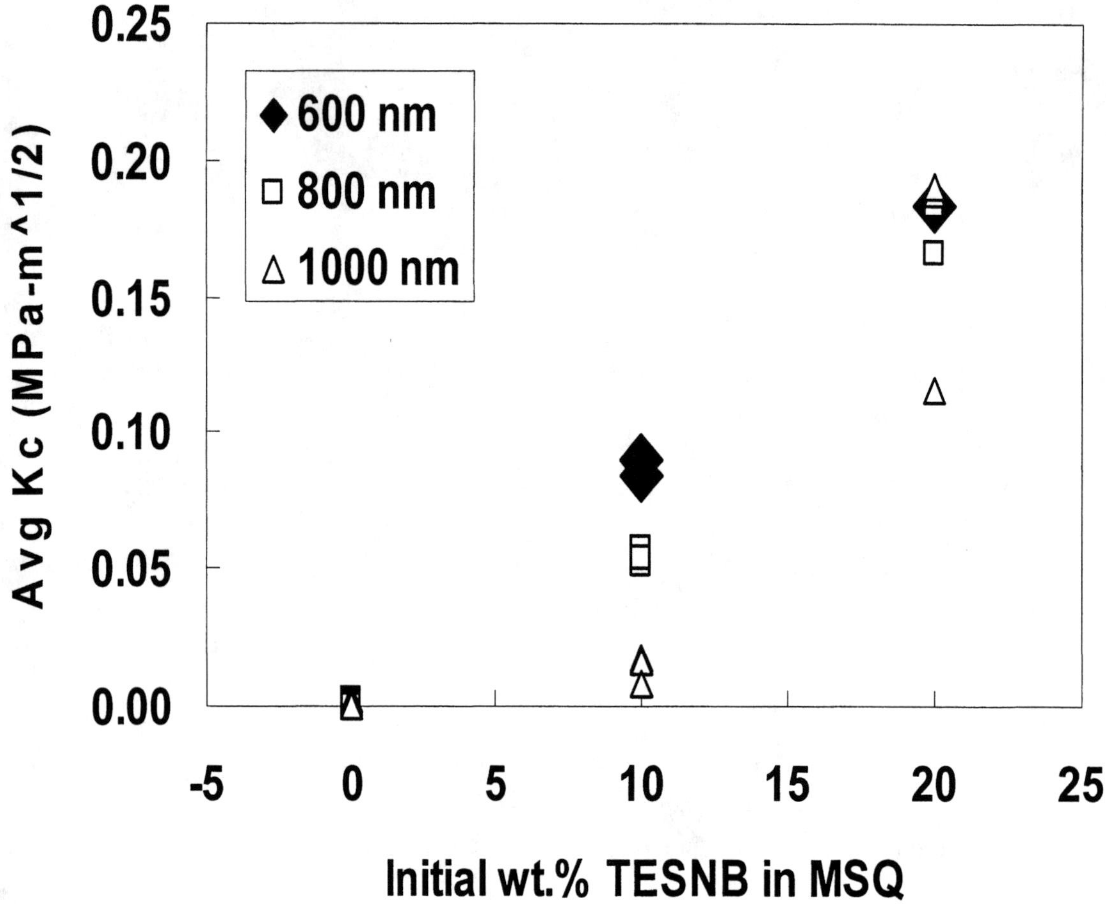

Figure 5. Fracture toughness of MSQ as a function of porosity.

Proceedings of
2001 ASME International Mechanical Engineering Congress and Exposition
November 11-16, 2001, New York, NY
EPP-Vol. 1

IMECE2001/EPP-24708

MICROTENSILE METHODOLOGY FOR MECHANICAL CHARACTERIZATION OF THIN FILMS

Betty H. Yeung*, Bill Lytle, Vijay Sarihan
Motorola Inc.
Tempe, Arizona 85284
Phone*: (480) 413-7734, Fax: (480) 413-4511
E-mail*: betty.yeung@motorola.com

David T. Read
National Institute of Standards and Technology
Boulder, Colorado 80303
E-mail: david.read@nist.gov

Yifan Guo
CiDRA Corporation
Wallingford, Connecticut 06492
E-mail: yguo@cidra.com

ABSTRACT

A microtensile methodology developed at the National Institute of Standards and Technology (NIST) has been adopted and applied at Motorola to evaluate material properties of thin films. This methodology is a significant part of the materials technology development at Motorola. Special specimens of thin metal films are designed and produced based on common microlithographic techniques and silicon processing methods. The experimentation is performed using the microtensile tester, which was developed for the accurate measurement of constitutive properties of thin metallic films. Through the application of the techniques presented here, valuable information and results have been achieved, which provide an extended information base for thin-film materials. Ultimately, such data are applied to processing and reliability predictions and the optimization of thin-film processes and materials.

INTRODUCTION

The application of thin films in semiconductor devices and packages is extensive and becoming increasingly important as new uses and more complex systems are being developed. The ability to better understand and more accurately represent thin-film structures is necessary. At the same time, to reduce design-cycle time, prediction of the behavior and the reliability of thin films in their given applications is becoming a critical necessity. The accuracy of these predictions is dependent on accurate determination of constitutive models of thin-films.

Understanding and evaluation in the area of thin films is becoming more widespread [1-3], but current materials characterization capabilities are common only for evaluation of bulk materials and are limited for thin films. It is well known that the behavioral response of thin films is significantly different from the characteristics of bulk materials of the same composition in properties such as strength, ductility, grain structure, and elongation. Their data are not able to accurately represent thin-film behavior. Unfortunately, the ability to handle and accurately determine the constitutive properties of thin films is a daunting challenge. To do so requires creating manageable specimens of the microstructure and geometry of thin films and accurately measuring their constitutive properties.

A microtensile testing methodology within Motorola is currently being utilized for the measurement of constitutive properties of thin films. The technology, developed at the National Institute of Standards and Technology (NIST) [4], involves the use of standard lithographic processes to produce damage-free thin films of a controlled geometry, a microloading system for the controlled application of tensile loads on the specimens, and the reduction of testing data to obtain constitutive properties. Recently evaluated materials include copper [5], aluminum, and nickel. The results for films of aluminum and nickel are presented here.

SPECIMEN FABRICATION

Key aspects involved in the fabrication of specimens include (1) the preparation of damage-free thin-film specimens in a controllable and reproducible manner, while (2) maintaining the features of real thin films in grain structure and geometry, and (3) creating suspended specimens with silicon acting as a rigid frame (Fig. 1). Up to 169 specimens can be produced on one 100 mm (4 in.) wafer.

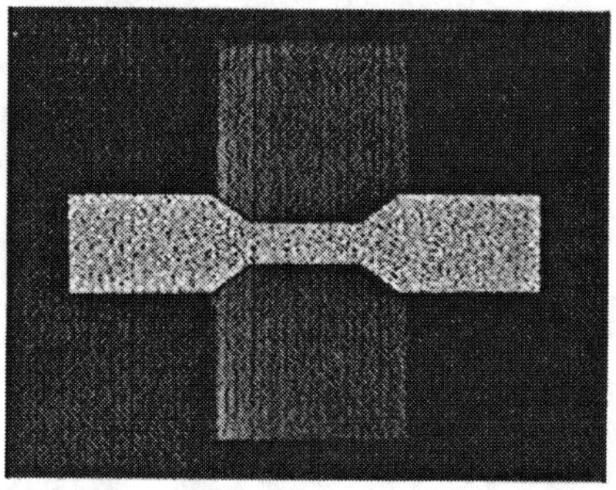

Figure 1. Thin-film test specimen

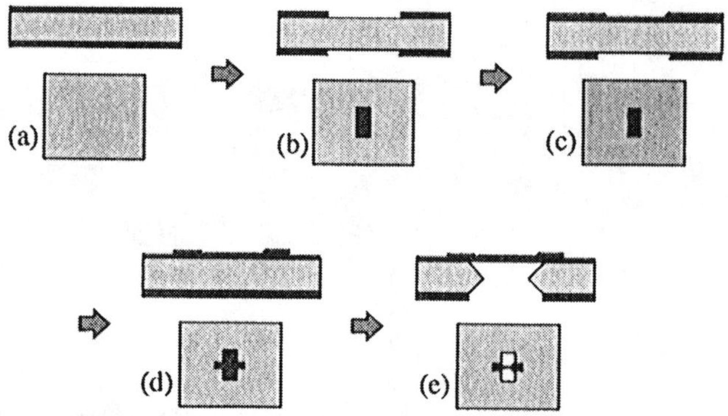

Figure 2. Procedures of thin-film specimen fabrication from transverse cross-sectional and plan views

Sample Generation

Common photolithographic techniques are applied to produce the tensile specimens. The series of steps involved in fabrication is depicted in Fig. 2. Transverse cross-sectional and plan views of the specimen are shown in the diagrams. The process begins with a silicon wafer of <100> orientation. A thin (~0.5 μm) layer of oxide is grown on both sides of the wafer (Fig. 2a), and is patterned with rectangular openings to leave "windows" of bare silicon (Fig. 2b). Narrow streets are also etched in the oxide during this step to aid in the subsequent manual singulation process. The metal(s) of interest is deposited over the entirety of one side of the wafer, ensuring uniform thickness and coverage of metal (Fig. 2c). Thus, a substantial sample size of specimens is generated for the given film thickness. The thin-film specimens, produced in alignment with the oxide windows, are formed when the metal is etched according to the mask-defined geometry (Fig. 2d).

Silicon Etching

The regions of bare silicon are removed using a unique silicon etching process that results in a freestanding thin-film specimen securely held by the silicon frame (Fig. 2e). This is the final step in the sample generation. The TMAH (tetramethyl ammonium hydroxide) silicon etching process, developed within Motorola, has been chosen to accomplish this. Prior to the silicon etch, the bare silicon surface is cleaned of any oxides that would prevent silicon from being etched. A solution of TMAH is maintained at a constant temperature during etching to achieve an optimum level of control and an etch rate of approximately 60 μm/h. Etching occurs from both sides of the wafer through the oxide windows. A thorough rinse ensures the removal of the etchant from the wafer following the etching process. The silicon etch creates cavities through the wafer, leaving a suspended thin-film specimen. For the thin-film metals that TMAH is known to attack, a means to protect the metal from the etchant is necessary.

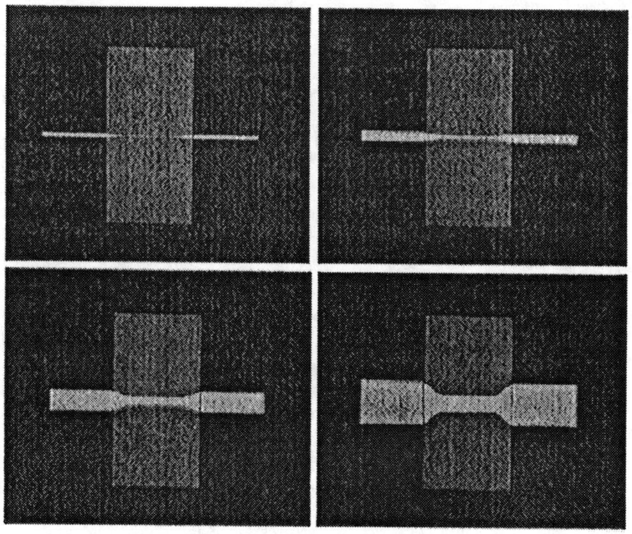

Figure 3. Typical thin-film specimen geometries

Specimen Geometry

Typical specimen geometries are shown in Fig. 3. The specimen gauge length is 600 μm, and the specimen width ranges from 20 μm to 200 μm. The total length of the specimen is about 2.5 mm. Film thicknesses evaluated to date range between 0.7 μm to 4 μm and are defined during processing. Actual geometries of the specimen, including length and width, are measured prior to testing. Actual thickness is determined from profilometric measurements.

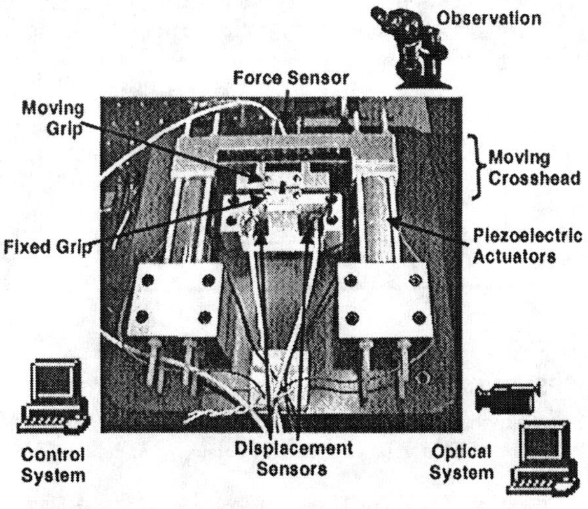

Figure 4. Schematic of microtensile test system

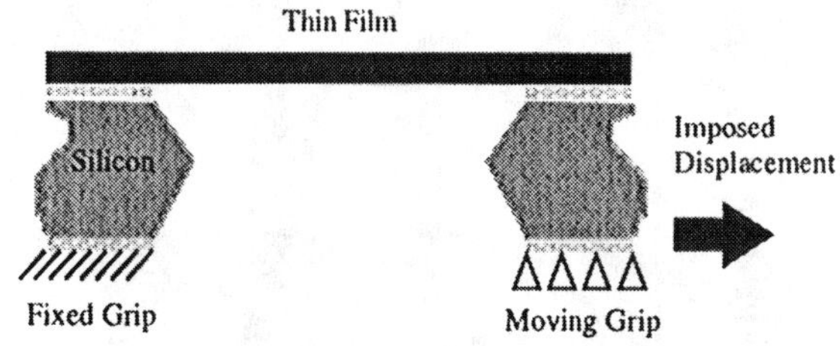

Figure 5. Microtensile loading of a thin-film specimen

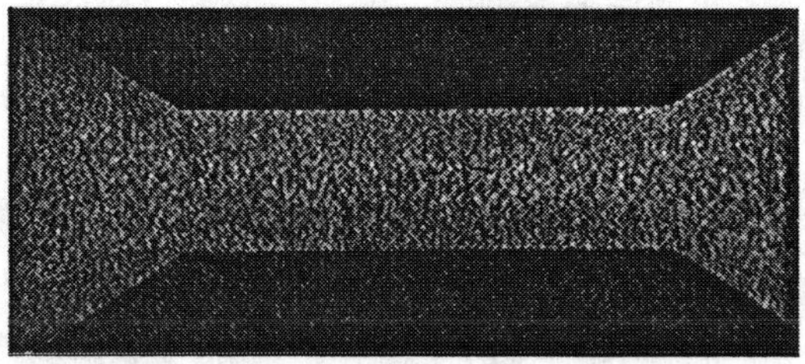

Figure 6. Typical specimen image used for strain analysis

TEST SYSTEM

The microtensile test system consists of a loading system and a computer control system. Figure 4 shows the complete assembly. The entire apparatus is mounted to a movable stage with an optical microscope for viewing capability during testing, specimen preparation, and specimen examination. Several distinct features of the test system include (1) the ability to apply forces in magnitudes and increments small enough to allow for controlled tensile displacements on the thin film, and (2) the capability to perform operations and preparations on the small specimen and frame without inducing premature damage or loading to the specimen. The main components of the test system include the fixed grip and the moving grip, which hold the tensile specimen, three sensors, two piezoelectric (PZT) stacks, a measuring beam, and a moving crosshead. The test system capability provides a displacement range of 50 μm ± 20 nm and a load range of 2 N ± 1 nN.

The two piezoelectric actuators parallel to the specimen length act as expandable columns. Through displacement control, a voltage is applied to expand the PZTs and displace the measuring beam. This action, in turn, induces a tensile load on the specimen. All moving components are cantilevered a small distance from the base of the device to eliminate any sliding frictional effects.

The load and displacement on the sample are measured as voltages through the non-contacting eddy-current sensors. The sensor voltages are converted through calibration factors to obtain actual displacement and applied force during the test. Two sensors located at the fixed grip measure the displacement of the moving grip, while a sensor mounted on the moving crosshead is used in conjunction with the measuring beam to monitor and record load.

EXPERIMENTAL APPROACH

Test Procedure

To prevent damage to the specimen during handling and preparation and to allow for removal and evaluation of the specimens after testing, a sample is attached to removable plates on the two grips using a cyanoacrylate adhesive. The specimen is secured at one end on the fixed grip and the other on the movable grip. The sides of the silicon frame are severed so that the specimen ends are free.

In its freestanding condition, the specimen is ready for testing. Typically, loading occurs at a constant ramp rate of displacement, though other loading profiles such as a sine wave or force control can be applied. A displacement is induced on the moving end of the specimen as shown in Fig. 5. The specimen is loaded the thin-film specimen fractures. Data acquisition during testing is performed through the computer control system using the custom NIGO software developed at NIST.

Image Analysis

To accurately obtain Young's modulus for thin films, two methods are employed. The 'apparent' modulus is obtained directly from the stress/strain data from the test. Since this modulus value is calculated from an overall elongation of the entire specimen, which includes the gauge length as well as the tapered ends, the resulting value is an approximation. Young's modulus is more precisely obtained through image-analysis techniques, including speckle interferometry [6] or digital image correlation [7-8]. Algorithms for the two methods developed at NIST allow for a defined gauge length with uniform cross-sectional area on the specimen to be tracked to achieve more reliable values of strain and, in turn, modulus. Figure 6 shows a specimen image used for digital image correlation analysis.

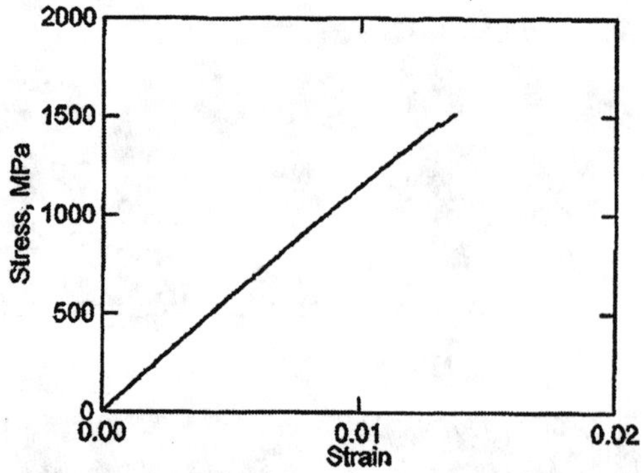

Figure 7. Tensile behavior for nickel thin film

RESULTS

The implementation of the methodology was successfully demonstrated for thin films of aluminum and nickel [9] produced within Motorola. The film's physical properties and the processing it underwent are significant factors affecting the resulting mechanical properties. From examination of the crystal structure, the electroless nickel is shown to be amorphous [10]. The nickel was plated to a thickness of ~4 μm isotropically on an aluminum film of ~0.7 μm thickness.

A typical stress-strain curve from a test of the nickel film is shown in Fig. 7. The stress-strain data indicate that the film is characterized by elastic behavior. The measured tensile strength of this thin film is higher than bulk values. This may be explained by the fact that bulk materials testing involves a relatively large volume through which numerous minute defects exist, while thin-film materials do not. Typical material properties, including ultimate tensile strength and Young's modulus were obtained for the nickel thin film and its underlayer, sputtered aluminum. The mechanical properties of the films obtained through the microtensile methodology are detailed in Table 1.

SUMMARY

Techniques to successfully evaluate constitutive properties of thin films are essential to efforts in both process and package development within Motorola and to the semiconductor industry in general. The use of standard thin-film preparation techniques allows for actual thin films that are used in current devices to be evaluated. The microtensile methodology incorporates the preparation and use of damage-free tensile specimens of thin films. The microtensile test system's unique capability to apply small-scale loads at appropriate levels allows for thin-film material properties to accurately be determined. This work demonstrates the successful application of techniques by which characteristic material properties can be obtained for thin films.

Table 1. Mechanical properties of thin films of aluminum and nickel measured using microtensile methodology

Thin Film Metal	Specimen Thickness [μm]	Specimen Width [μm]	Ultimate Tensile Strength [MPa]	Elongation to Fracture [μm]	Young's Modulus (Apparent) [GPa]	Young's Modulus (Image Analysis) [GPa]
Aluminum	0.73	196	469	11	45.9	
		194	457		44.4	
		45	589		40.8	
Nickel	4.69	99	1512	14	114	102
		100	1520		82	102

REFERENCES

[1] Alexopoulos, P. S. and O'Sullivan, T. C., "Mechanical Properties of Thin Films", *Annual Reviews of Material Science* **20**, 1990, pp. 391-420.

[2] Long, G. S., Read, D. T., McColskey, J. D., and Crago, K, Microstructural and Mechanical Characterization of Electrodeposited Gold Films, *ASTM STP 1413*, 2001, *Mechanical Properties of Structural Film.*

[3] Hardwick, D. A., "The Mechanical Properties of Thin Films, A Review," *Thin Solid Films* **154**, 1987, pp. 109-124.

[4] Read, D. T., "Piezo-Actuated Microtensile Apparatus," *Journal of Testing and Evaluation*, JTEVA, **26** (3), May 1998, pp. 255-259.

[5] Volinsky, A. A., Vella, J., Adhihetty, I., Sarihan, V., Mercado, L., Yeung, B., and Gerberich, W., "Microstructure and Mechanical Properties of Electroplated Copper Thin Films", *MRS Symposium Q Symposium Proceedings*, Fall 2000.

[6] Read, D. T., "Young's Modulus of Thin Films by Speckle Interferometry," *Measurement Science and Technology 9*, 1998, pp. 676-685.

[7] Peters, W. H., "Digital Imaging Techniques in Experimental Stress Analysis", *Optical Engineering*, **2** (3), 1982, pp. 427-431.

[8] Sutton, M. A., McNeill, S. R., Helm, J. D., Schreier, H., and Chao, Y. J., "Advances in 2D and 3D Computer Vision for Shape and Deformation Measurements", *Photomechanics*, P. K. Rastogi, Ed., Topics in Applied Physics, Springer Verlag, 77, 2000, pp. 323-372.

[9] Grupen-Shemansky, M., Thompson. P, Lin, J. K., "Alternate Alloy Bump Interconnect Technologies and Flip Chip Applications", *ASME Advances in Electronic Packaging*, EEP-Vol. 26-2, Vol. 2, 1999, pp. 1303-1308.

[10] Jang, J. W., Kim, P. G., Tu, K. N., "Low Cost Flip Chip Technology: Solder Reaction-assisted Crystallization of Electroless Ni-P Under Bump Under Bump Metallization", *6th Report to SEMATECH from UCLA*, Sept. 1998.

Proceedings of
2001 ASME International Mechanical Engineering Congress and Exposition
November 11-16, 2001, New York, NY
EPP-Vol. 1

IMECE2001/EPP-24709

MECHANICS ISSUES AT MICRO-SCALE MODELING OF ELECTRONIC PACKAGES

Raghuram V. Pucha, Gnyaneshwar Ramakrishna, and Suresh K. Sitaraman*

Computer Aided Simulation for Packaging Reliability Lab
The George W. Woodruff School of Mechanical Engineering
Georgia Institute of Technology
Atlanta GA 30332-0405
*Voice: 404-894-3405
FAX: 404-894-9342
suresh.sitaraman@me.gatech.edu

ABSTRACT

Due to the temperature dependent material properties and low cycle thermo-mechanical fatigue of electronic packaging interconnects and components, plasticity theories are extensively used to understand their mechanical response in estimating the fatigue life. Experimental evidence indicates that plastic deformation at small scales depends not only on the state variables of stress and strain, but also on their higher order gradients. In this paper mechanics issues related to strain gradient plasticity at micron scale are reviewed through strain gradient enhanced continuum theories. A plastic strain gradient-based approach is employed to study the thermo-mechanical deformation of microvia structure.

INTRODUCTION

The continued push for ultra fine-pitch, increased functionality and improved performance in semiconductor devices necessitates innovative interconnect technology with reduced feature sizes in the range of few microns to meet the International Technology Roadmap for Semiconductor (ITRS) packaging requirements for year 2012 and beyond. Sometime in the future such chip to package/substrate interconnects may be of sub-micron size.

Electronic packaging assemblies consist of components having dimensions that are of different length scales. Therefore, a modeling approach needs to address material behavior taking into consideration their different sizes. For example, the modeling may deal with solder dimensions of tens of microns to substrate dimensions of 150 X 150 mm with microstructure dependent material properties. Efforts toward innovative interconnect technology may lead to feature dimensions of the order of microns or sub-microns. The conventional continuum formulations may fail at some length scales, as the mechanics issues associated with various length scales are different. For example in mesoscale continuum mechanics size matters [1]. The critical deformation and fracture process in structural metals for example happens at mesoscale. The mesoscale formulations are intermediate between a direct atomistic and an unstructured continuum description of deformation process. Dealing with mesoscale mechanics may be too small for continuum mechanics and too large for direct atomistic mechanics.

Some examples of design and manufacturing applications at micron level includes thin films where film thickness is on the order of 1 micron or less; sensors, actuators and MEMS devices where the entire system size is less than 10 microns; and microelectronic packages where features such as vias could be smaller than 15 microns. The material length scale can have effect on the material properties, elastic deformations or plastic deformations. This paper is concerned with material length scale effects on plastic deformation.

Materials display strong scale effects when the characteristic length scale associated with non-uniform plastic deformation is on the order of microns. The classical plasticity theories cannot predict the size dependence of material behavior at the micron scale because their constitutive models typically possess no internal length scale. There is accumulated experimental evidence for the existence of material size effects in plasticity [2]. In particulate reinforced metal matrix composites, small particles give rise to an enhanced rate of strain hardening compared to the same volume fraction of larger particles [3,4]. Torsion tests on copper wires show that the scaled shear strength increases by a factor of three as the wire diameter decreases from 170 μm to 12 μm, while the increase of work-hardening in simple tension is negligible [5]. In the bending of ultra thin beams, Stolken and Evans [6] observed a significant increase in the normalized bending hardening as the beam thickness decreases from 100 μm to 12.5 μm, while the data from simple tension displays no size dependence. Lloyd [7] observed a substantial strength increase when the particle diameter was reduced from 16 μm to 7.5 μm with the particle volume fraction fixed at 15%. Indentation tests show that inferred hardness increases with diminishing indent size for indents in the micron to sub-micron range. Large strain gradients inherent in small indentations lead to geometrically necessary dislocations that cause enhanced hardening. The measured indentation hardness of metallic materials was observed to increase by a factor of two as the depth of indentation decreases from 10 μm to 1 μm [8-13].

Experimental evidence, from above mentioned references, indicates that plastic deformation at small scales depends not only on the stress and strain, but also on their higher order gradients, whereas in conventional theories of plasticity, the flow stress is uniquely related to the strain at that point and is unrelated to any strain gradients that might be present. Gradient-enhanced continuum mechanics theories are applicable when the component size is comparable to some internal length scale of the material, typically on the order of microns. This paper presents a brief review of mechanics issues related to strain gradient plasticity at micron scale. A typical case study from packaging applications is presented to demonstrate the scale effect due to strain gradient plasticity and its influence on the thermo-mechanical response.

REVIEW OF STRAIN GRADIENT PLASTICITY THEORIES

Most of the higher-order continuum mechanics theories include not only stresses and strains in the constitutive equations, but also invariants of their derivatives, or additional derivatives of displacements. Fleck and Hutchinson [14,15] have developed a phenomenological theory of strain gradient plasticity based on macroscopically measured uniaxial stress-strain curve. This theory fits the mathematical framework of higher-order continuum theories of elasticity, with the strain gradients represented either in terms of the gradients of rotation or in terms of both rotation and stretch gradients. The theory was formulated by replacing effective stresses and strains in conventional plasticity with higher-order effective stresses and strains. The higher-order stresses and strains contain gradient terms scaled by a phenomenological material length to be determined from experiments and the framework is within the context of isotropic hardening plasticity theory.

Strain hardening of materials is due to the accumulation of dislocations. When a material is deformed, dislocations are generated, moved and stored, and the storage causes the material to work harden. Dislocations become stored for one of the following two reasons: (1) Dislocations accumulate by trapping each other in a random way, which are known as statistically stored dislocations. (2) When dislocations are required for compatible deformation of various parts of the material they are known as geometrically necessary dislocations. Statistically stored dislocations are created by homogeneous strain while geometrically necessary dislocations are related to the strain gradients. Both dislocations contribute to the flow stress. The degree of hardening due to statistically stored dislocations is assumed to scale with the von Mises effective strain. The hardening due to geometrically necessary dislocations is taken to scale with an isotropic scalar measure of the strain gradient in the deformed solid, and with a material length parameter l. Using this multi-scale hierarchical framework relating microscale dislocations to mesoscale plasticity, a mechanism-based theory of strain gradient plasticity theory was proposed by Gao et al [16]. This hierarchical structure provides an approach for constructing the mesoscale constitutive laws by averaging microscale plasticity laws over the representative cell. This mechanism-based theory was used to describe the elastic-plastic instabilities and experimentally observed plasticity scale effects in bending of thin beams and torsion of thin wires [17].

All the above mentioned strain gradient theories introduce the higher-order stress as work conjugate of strain gradients, which is required to satisfy thermodynamic restrictions on the constitutive model for second deformation gradients. Alternative gradient theories without work conjugate of strain gradients are also available in the literature. Representing the strain gradient effects as the first and second Laplacian of effective strain, Aifantis [18] modified the constitutive model of classical plasticity (yield function) by introducing the Laplacian of plastic strain in the relation between flow stress and plastic strain. Acharya and Bassani [19] proposed a strain gradient plasticity theory, which retains the essential structure of conventional plasticity and obey the thermodynamic restrictions on the constitutive relations. Their formulation is based on flow theory with strain gradient effects represented as an internal variable, which acts to increase the current tangent-hardening modulus. The strain gradients only appear implicitly in the constitutive relation through the incremental plastic moduli and the density of geometrically necessary dislocations is incorporated into the hardening description. However, there has not been a systematic way of constructing the tangent modulus to validate this approach. In a recent work, Chen and Wang [20] proposed a new hardening law for strain gradient plasticity, retaining the essential structure of conventional plasticity theory, in which the strain gradient effect is introduced as an internal variable which acts to increase the current tangent modulus in defining the new hardening law. Though not yet used extensively in the field of microelectronics, gradient-enhanced continuum mechanics theories have been applied and found useful for analyses of many micrometer scale problems such as (i) asymptotic crack tip fields in interface fracture problems [21,22] (ii) thin-film delamination problems [23] and (iii) strain localization problems in plasticity [18].

Finite element implementations of strain gradient theories described above are useful, while standard continuum mechanics will not capture important gradient effects and at the same time atomistic modeling may be impractical. Investigation into the results of including strain-gradient effects in a finite element analysis will require special purpose finite elements and models. Inclusion of strain gradient effects in finite element analysis for example requires the implementation of elements with (i) higher-order degrees of freedom, with displacement gradients as nodal degrees of freedom and either C^1 continuous elements or C^0 elements with Lagrange multipliers to enforce kinematic constraints on displacement gradients, (ii) increased inter-element continuity as well as (iii) new constitutive laws that include strain and stress gradient terms.

Finite element implementations of both Laplacian and first-order strain gradient theories have been developed. De Borst and Pamin [24] developed families of elements for Laplacian strain gradient theory, one set using C^0 continuous elements, and the other using C^1 continuous elements. Li and Cesotto [25] developed C^0 elements, where the Laplacian of equivalent of equivalent plastic strain is calculated from a polynomial interpolation through adjacent integration points. The yield condition is no longer local, but is satisfied in a point-wise fashion at each iteration of a load step. Shu, King, and Fleck [26] developed C^0 continuous elements for the first-order strain gradient theory. These elements use displacements and displacement gradients as nodal degrees of freedom. The Lagrange multiplier method is used to enforce kinematic constraints between displacement gradients. In each of these works, the authors have been able to show good agreement with analytical predictions for simple cases.

OBJECTIVE OF THE PRESENT WORK

Due to the temperature dependent material properties and low cycle thermo-mechanical fatigue of electronic packaging interconnects and components, plasticity theories are extensively used to understand their mechanical response in estimating the fatigue life. The established algorithms in conventional computational plasticity cannot be used for higher-order strain gradient plasticity theories, and on the other hand, a significant effort is required to develop special purpose finite elements for strain gradient plasticity theories in the analysis of microelectronic packaging applications. The objective of this paper is to explore the possibilities of introducing the strain gradient effects in commercially available displacement-based finite element codes by iteratively updating the material behavior in the strain gradient influenced plastic zone. A typical case study from packaging applications is presented to demonstrate the feasibility of such an approach in understanding the scale effects due to gradient plasticity using strain gradient plasticity theories. The potential influence of gradient effects on the flow stress using Gao et al. strain gradient theory [16] is used to update the material hardening relation in the strain gradient influenced plastic zone. A model of copper microvia thermo-mechanical reliability is used to demonstrate the influence of strain gradient effects on plastic strain evolution using Cheng and Wang [20] approach of iteratively updating the material hardening in the incremental stress strain relation.

MICROVIA RELIABILITY - PLASTIC STRAIN GRADIENT APPROACH

Microvias play a key role in electronic packaging to realize high density interconnects on the printed wiring board at low cost. As the demand to increase the wiring density of the various layers on the board increases, there is an increasing need to reduce the size of the microvias. Presently the smallest microvia size that is fabricated for commercial application has a 50-micron diameter. The demand is to fabricate sub mil (<25 microns) vias in coming years. The wall plating thickness is the smallest dimension in the geometry of microvia. As the diameter of the vias reduces, the wall thickness correspondingly reduces and hence the minimum feature size reduces to the order of microns, where scale effects in plasticity are predominant. In predicting the fatigue life of these microvias accurate estimation of plastic strain evolution and range is critical. Microstructure dependent plastic deformation analysis incorporating strain gradient plasticity theories are attractive in this context. The thermo-mechanical reliability analysis of microvia structure is presented here to demonstrate the influence of incorporating strain gradient effects in predicting the evolution of plastic deformation.

A single sided multi-layered high-density wiring structure is modeled with FR4 as the base substrate. The substrate has a single layer of dielectric in which the microvias are located. A schematic of the geometry is shown in Figure 1(a). A plane strain finite element model of microvias with 20 μm inner diameter with a plating thickness of 4 μm and an outer diameter of 40μm is considered as shown in Figure 1(b). The FR4 is modeled as an orthotropic material, while the dielectric is modeled as a viscoelastic material. The copper microvias are modeled using a multi-linear isotropic elasto-plastic model. Stresses and strains in the microvia are determined by simulating the cooling down of the structure from 165° C to -55° C in six minutes.

Figure 2 shows the variation of effective plastic strain distribution in the microvia at the end of the cooling at -55° C. Upon cooling the structure warps due to the presence of mismatch in the coefficient of thermal expansion (CTE) of dielectric on top of FR4. Due to the thermo-mechanically induced extension and warpage, plastic strain localization occurs at the wall plating junctions of the microvia as expected. Following the equations described in Appendix 1, the corresponding strain gradients are estimated and are shown in Figure 3. The region A in Figure 3 corresponds to material with practically no change in hardening behavior with small strain gradients. In regions B and C the gradients are as large as 140/mm and the material hardens more. The flow stress ratio, which is defined as the ratio of flow stress with and without strain gradient effects, estimated as described in Appendix 1, is used to define the updated material hardening due to the strain gradients.

An iterative algorithm for strain gradient modified material hardening, with following steps, is proposed for incorporating the strain gradient dependent plasticity theory without resorting to special purpose finite elements. The algorithm is demonstrated with the microvia thermo-mechanical analysis by implementing them in a commercially available finite element code to study the effect of strain gradient dependent hardening relation on plastic deformation.

Step1: Fit the material data from the initial stress – strain curve $\sigma = K\ \varepsilon^N$ to get hardening modulus, K and strain hardening exponent N.

Step 2: Using the strain distribution from the finite element analysis compute the effective strain gradients and flow stress ratio for each element, as shown in Appendix 1.

Step 3: The strain gradient modified hardening modulus for each element is calculated as

$$K_{new} = K_{previous} * \text{Flow stress ratio}$$

Step 4: Each element material hardening relation is modified in the subsequent analysis with K_{new}

Step 5: Repeat steps 2-4 until convergence with respect to decrease in the strain gradients occurs.

By iteratively changing the stress-strain curve (hardening relation), incorporating the strain gradient effects back into the analysis as described above, Figure 4 shows the snap shots of strain gradient distribution at the end of iterations 2,4,6 and 8. It is to be noted here that in the areas where the gradient is small (region A in Figure 3), the K_{new} does not change. The material hardening due to plastic strain gradients and the corresponding value of maximum strain gradient in the successive iterations are shown in Figure 5(a) and 5(b) respectively. The increasing material hardness in the successive iterations due to incorporating strain gradients back into the analysis (Figure 5(a)), results in convergence with respect to reduced strain gradient effects in the successive iterations (Figure 5(b)), leading to uniform strains at the end of ten iterations as shown in Figure 6(a). Figure 6(b) shows the variation of plastic strain gradients at the end of ten iterations. An expected trend towards reduced strain gradients indicates that strain gradient modified hardening relation dissipates the strain localization and predicts the plastic strain distribution due to the true material behavior in view of the scale effects. It has important consequences in the predicting the fatigue life of these structures based on the plastic strain distribution. It is expected that the reduced feature sizes of the order of microns can survive more cycles when strain gradient enhanced plasticity theories are used.

CONCLUSIONS

Classical plasticity theories are extensively used in the plastic strain-based fatigue analysis for thermo-mechanical reliability of electronic packaging structures. With the increased density and reduced feature sizes the classical plasticity theories may fail at some length scales when the feature dimensions are on the order of microns. With the increasing experimental evidence of scale effects in plasticity due to strain gradients and the development of strain gradient plasticity theories, there is a need for developing corresponding special purpose finite element models. This paper is an attempt towards exploring the possibility of developing a computational algorithm without resorting to special purpose finite elements for strain gradient plasticity. A case study from electronic packaging applications is presented to demonstrate the scale effects in plasticity and the effect of strain gradient modified material hardening on the plastic strain evolution. Reduced strain gradients are observed in the successive iterations due to the incorporation of strain gradient dependent hardening relation.The presented results on plastic strain distribution with and without considering the strain gradient plasticity are expected to have consequences (i) in understanding the mesh dependent plastic strain localization and (ii) in predicting the life of packaging structures when the feature sizes are of the order of microns. The plastic strain distributions presented in this paper, by incorporating strain gradient-based plasticity approach, need to be validated with Moiré-based deformation data.

REFERENCES

1. Needleman, A., 2000, "Computational Mechanics at the Mesoscale", *Acta Materialia*, 48, 105-124.
2. Hutchinson, J.W., 2000, "Plasticity at Micron Scale", *International Journal of Solids and Structures,* 37, 225-238.
3. Kelly, A., and Nicholson, R.B., 1963, "Precipitation hardening", *Prog. Mater. Sci.* 10(2), 149-391.
4. Ebeling, R., and Ashby, M.F., 1966, "Dispersion hardening of copper single crystals", *Phil. Mag.,* 13, 805-834
5. Fleck, N.A., Muller, G.M., Ashby, M.F., Hutchinson, J.W., 1994, "Strain gradient plasticity: theory and experiment", *Acta Metallurgica et Materialia*, 42, 475-487.
6. Stolken, J.S., and Evans, A.G., 1998, "A microbond test method for measuring the plasticity length scale", *Acta Materialia*, 46, 5109-5115.
7. Lloyd, D.J., 1994, "Particle reinforced aluminum and magnesium matrix composites", *Int. Mater. Rev*, 39. 1-23.
8. Nix. W.D., 1989, "Mechanical properties thin films", *Metall. Trans.* 20A. 2217-2245.
9. De Guzman, M.S., Newbaur, G., Flinn, P., and Nix, W.D., 1993, "The role of indentation depth on the measured hardness of materials", *Materials research proceedings*, 308, 613-618.
10. Stelmeshenko, N.A., Walls, M.G., Brown, L.M., and Milman, Y.V., 1993, "Microindentation on W and Mo oriented single crystals; an STM study", *Acta Metallurgica et Materialia*, 41, 2855-2865.
11. Ma. Q., and Clarke, D.R., 1995, "Size dependent hardness in silver single crystals", *J. Materials Research*, 10. 853-863.
12. Poole, W.J., Ashby, M.F., and Fleck, N.A., 1996, "Microhardness of annealed and work-hardened copper polycrystals" *Scripta Metall. Mater*, 34, 559-564.
13. McElhaney, K.W., Vlassak, J.J., and Nix, W.D., 1998, "Determination of indenter tip geometry and indentation contact area for depth-sensing indentation experiments", *J.Materials Research,* 13, 1300-1306.
14. Fleck, N.A., and Hutchinson, J.W., 1993, "A phenomenological theory for

strain gradient effects in plasticity" *J.Mech. Phys. Solids,* 41, 1825-1857.
15. Fleck, N.A., and Hutchinson, J.W., 1997, "Strain gradient plasticity". In: Hutchinson, J.W., Wu, T.Y. (Eds.) *Advances in Applied Mechanics*, Vol.33. Academic Press, New York, 295-361.
16. Gao, H., Huang, Y., Nix, W.D., and Hutchinson, J.W., 1999, "Mechanism-based Strain gradient plasticity – I.Theory", *Journal of Mechanics and Physics of Solids*, 47, 1239-1263.
17. Huang, Y., Gao, H., Nix, W.D., and Hutchinson, J.W., 2000, "Mechanism-based strain gradient plasticity – II. Analysis", *Journal of mechanics and physics of solids*, 48, 99-128.
18. Aifantis, E.C., 1992, "On the role of gradients in the localization of deformation and fracture", *Int. J. Engng Sci.*, 30, 1279-1299.
19. Acharaya. A., and Bassani, J.L., 1995,"On non-local theories that preserve the classical structure of incremental boundary value problems". In: Micromechanics of plasticity and damage of multiphase materials. *IUTAM Symposium,* 3-9.
20. Chen, S.H., Wang, T.C., 2000, "A new hardening law for strain gradient plasticity", *Acta Materialia*, 48, 3997-4005.
21. Xia, Z.C., and Hutchinson, J.W., 1996, "Crack tip fields in strain gradient plasticity" *J. Mech. Phys. of Solids*, 44, 1621-1648.
22. Haung, Y., Zhang, L., Guo, T.F., and Hwang, K.-C., 1997, "Mixed mode near tip fields for cracks in materials with strain-gradient effects", *J. Mech. Phys. of Solids*, 45, 439-465.
23. Wei, Y., and Hutchinson, J.W., 1999, "Models of interface separation accompanied by plastic dissipation at multiple scales", *Int. J. Fracture*, 95, 1-17.
24. De Borst, R., and Pamin, J., 1996, "Some Novel Developments in Finite Element Procedures for Gradient-Dependent Plasticity", *Int. J. Numerical Meth. Engng.*, 39, 2477-2505.
25. Li, X., and Cescotto, S., 1996, "Finite Element Method for Gradient Plasticity at Large Strains", *Int. J. Numerical Meth. Engng.*, 39, 619-633.
26. Shu, J., King, W., and Fleck, N. A., 1999, "Finite Elements for Materials with Strain Gradient Effects", *Int. J. Numerical Meth. Engng.*, 44, 373-391.

APPENDIX 1

Estimation of effective strain gradients

From the continuum plasticity point of view the density of geometrically necessary dislocations are linked to the deformation curvature, which can be represented by a third order strain gradient tensor

$$\eta_{ijk} \equiv u_{k,ij}$$

Where u_k are displacements from the finite element analysis.

η_{ijk} are computed at centriod of each element and the effective strain gradient is defined as

$\eta = \sqrt{\eta_{ijk}\eta_{ijk}}$ and are computed to estimate the flow stress ratio for the each element as follows.

Flow stress with and without strain gradients

The Taylor relation between shear strength and dislocation density in a material is given by

$$\tau = \alpha \mu b \sqrt{\rho_T} = \alpha \mu b \sqrt{\rho_S + \rho_G}$$

Where
ρ_T = Total dislocation density
ρ_S= Density of statistically stored dislocations
ρ_G = Density of geometrically necessary dislocations
μ= Shear modulus
b = Burgers vector
α = An empirical constant which depends on the crystal type

The degree of hardening due to statistically stored dislocations is assumed to scale with the von Mises effective strain, while the geometrically necessary dislocations are related to the gradients in the strain field in a material.

The effective strain gradient $\eta = \rho_G b$

Following Gao et al. strain gradient plasticity theory [16], in the case of the absence of strain gradient terms the flow stress is related to the uniaxial power law stress strain relation

$$\sigma = \sigma_{ref}\varepsilon^N$$

The theory proposes a modified flow stress, which depends on the internal material length scale and the effective strain gradient as

$$\sigma = \sigma_{ref}\sqrt{\varepsilon^{2N} + \ell\eta}$$

The expression for the factor increase in the flow stress (flow stress ratio) due to strain gradient effects is given by

$$\text{Flow stress ratio} = \frac{\sqrt{\varepsilon^{2N} + l\eta}}{\varepsilon^{N}}$$

Where σ_{ref} = Reference stress
N = Strain hardening exponent
l = Internal length scale
η = Effective strain gradient

The internal length scale is on the order of microns and l = 2.8 microns is considered for copper material [20] in the case study presented.

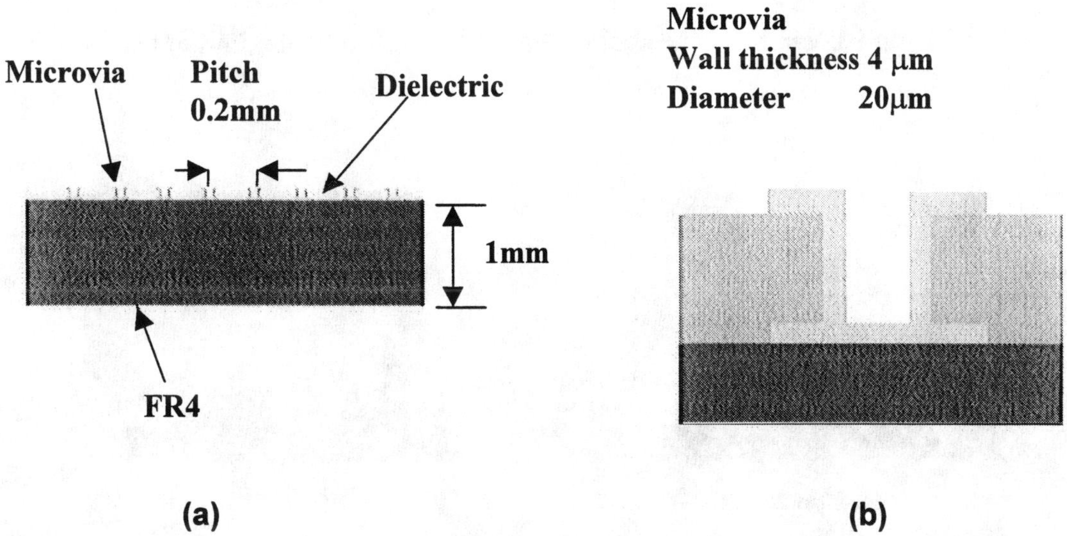

Figure 1. Schematic of the Microvia Geometry

Figure 2. Variation of Effective Plastic Strain (End of cooling)

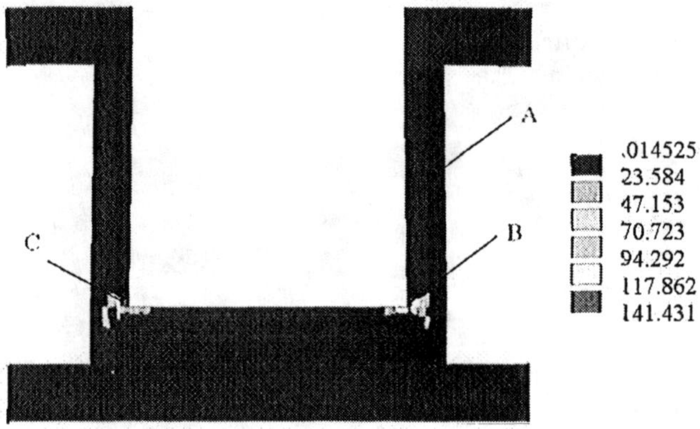

Figure 3. Variation of Effective Plastic Strain Gradients (End of cooling)

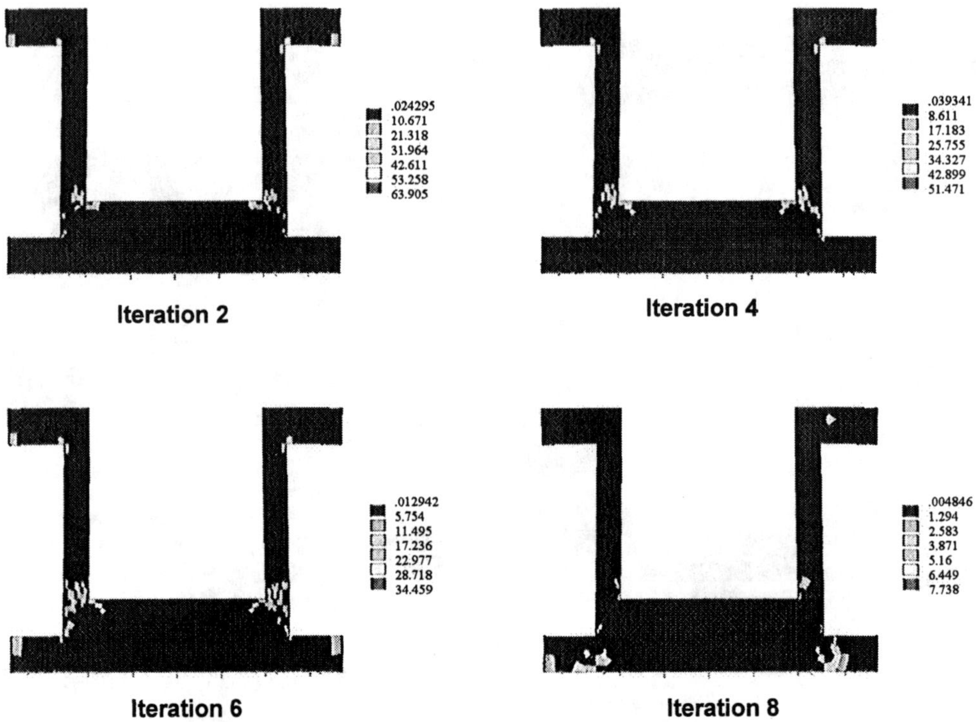

Figure 4. Snap shots of Effective Plastic Strain Gradients (Iteration 2,4,6 and 8)

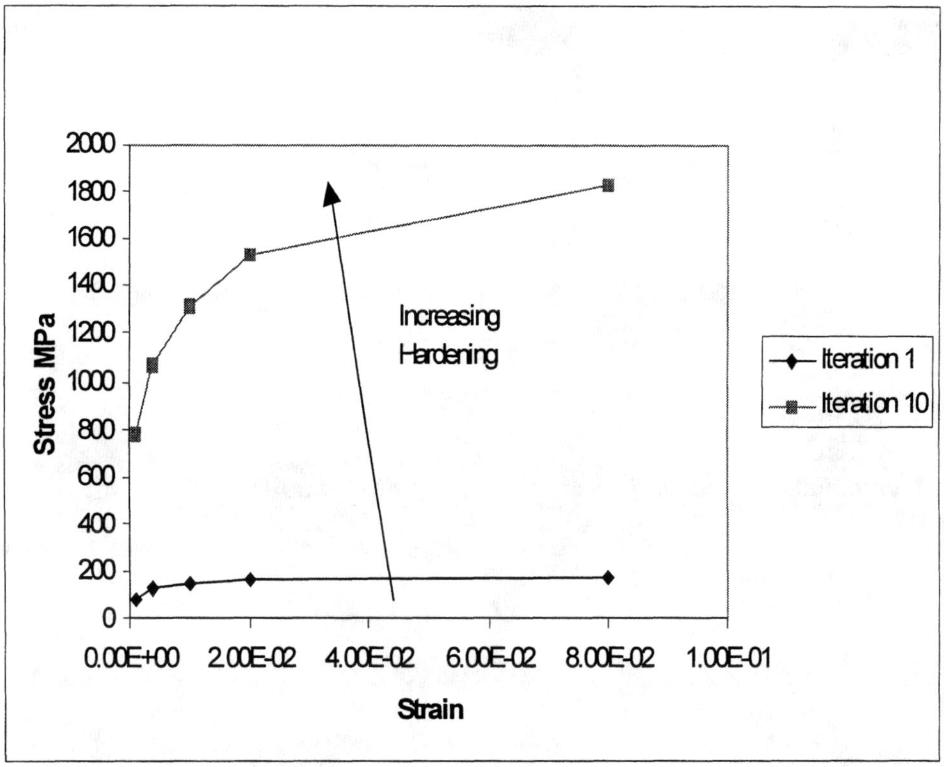

Figure 5(a). Variation Plastic Strain Gradient Modified Hardening

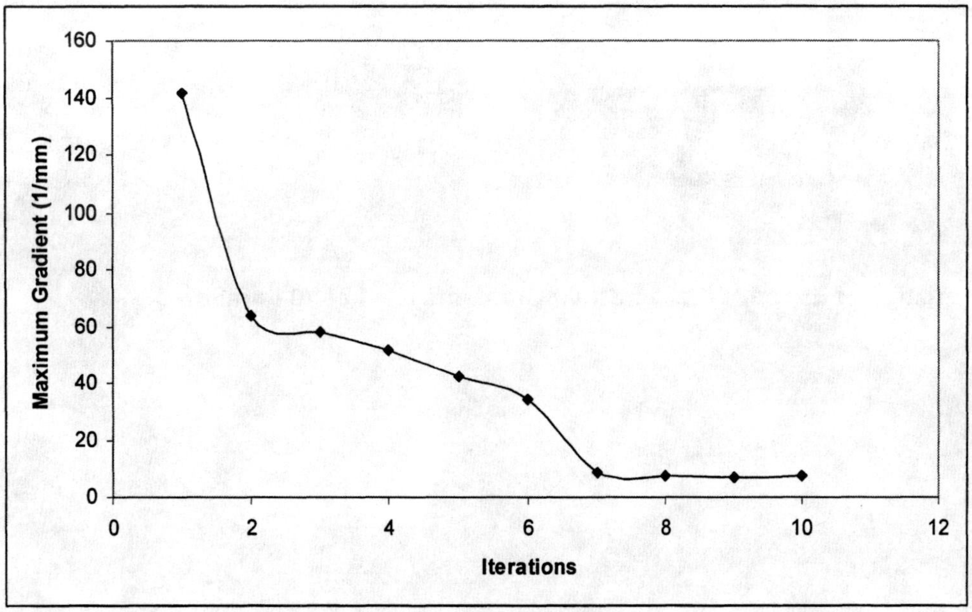

Figure 5(b). Variation of Maximum Plastic Strain Gradient

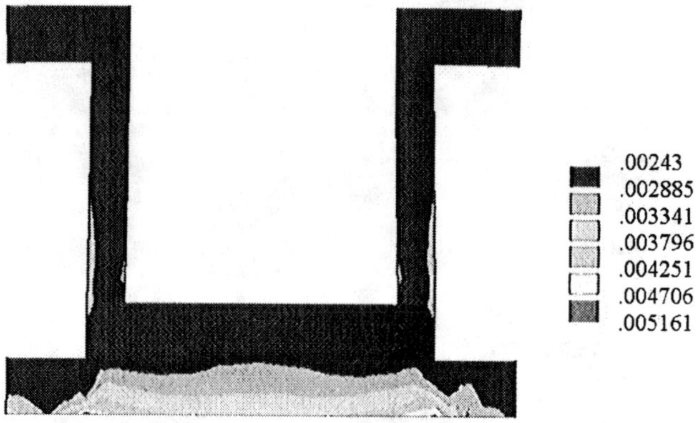

Figure 6(a). Variation of Effective Plastic Strain (End of 10 Iterations)

Figure 6(b). Variation of Effective Plastic Strain Gradients (End of 10 Iterations)

Proceedings of
2001 ASME International Mechanical Engineering Congress and Exposition
November 11-16, 2001, New York, NY
EPP-Vol. 1

IMECE2001/EPP-24710

MINIATURE VAPOR COMPRESSION REFRIGERATION SYSTEMS FOR ACTIVE COOLING OF HIGH PERFORMANCE COMPUTERS

Ali Heydari
Senior Thermal Engineer, RASCAL, CTO
Sun Microsystems
901 San Antonio Rd., UMPK15-222
Palo Alto, CA 94303
Ali.Heydari@sun.com

Kathy Russell
Thermal Engineer, RASCAL, CTO
Sun Microsystems
500 Eldorado Blvd., BRM05-108
Broomfield, CO 80021
Kathy.Russell@sun.com

ABSTRACT

A small refrigeration system for cooling of computer system components is evaluated. A thermodynamic model describing the performance of the cycle along with a computer simulation program is developed to evaluate its performance. The refrigeration system makes use of a miniature reciprocating vapor compression compressor. Due to space limitations in some high performance computer servers, a miniature refrigeration system composed of a compressor, capillary tube, a compact condenser, and a cold-plate evaporator heat exchanger are used. Mathematical multi-zone formulation for modeling thermal-hydraulic performance of heat exchanger for the condenser and evaporator are presented. The throttling device is a capillary tube and there is presented a mathematical formulation for predicting refrigerant mass flow rate through the throttling device. A physically based efficiency formulation for simulating the performance of the miniature compressor is used. An efficient iterative numerical scheme with allowance for utilization of various refrigerants is developed to solve the governing system of equations. Using the simulation program, the effects of parameters such as the choice of working refrigerant, evaporating and condensing temperatures on system components and overall efficiency of system are studied. In addition, a RAS (reliability, availability and serviceability) discussion of the proposed CPU-cooling refrigeration solution is presented. The results of analysis show that the new technology not only overcomes many shortcomings of the traditional fan-cooled systems, but also has the capacity of increasing the cooling system's coefficient of performance.

NOMENCLATURE

Q = Heat transfer rate (W)
T = Refrigerant temperature (^oC)
t = Air temperature (^oC)
U = Overall heat transfer coefficient ($W/m^{2\,o}C$)
i = Enthalpy (kj/kg)
h = heat transfer coefficient ($W/m^{2\,o}C$)
Δt = Temperature difference (^oC)
$m^{\cdot}$ = Mass flow rate (kg/hr)
W = Flow rate (kg/hr)

ensity (kg/ m^3)

σ = Surface tension (N/m)
μ = Dynamic viscosity (Ns/m^2)
v = Specific volume (m^3/kg)
C_p = Specific heat at constant pressure (kJ/kg ^{o}C)
C = Evaporator pressure regulator constant
L = Tube length (m)
l = Thickness of the cold plate wall (m)
K = Thermal conductivity (W/m. ^{o}C)
V' = Air volumetric velocity (m^3/hr)
F = Fraction of length of tube length
P = Pressure (kPa)
ΔP = Pressure difference (kPa)
A = Flow cross sectional area (m^2)
d = Tube diameter (m)
W = Work (Watts)
P = Power input (hp)
ANNL = Length fraction of the tube with quality up to 0.85
XDRY = Length fraction of the tube with quality within the range of 0.85 to 1.0
b = Correlation coefficient
C = Correlation coefficient
π_1= Geometry effect = L/d
π_2= Vaporization effect = $d^2h_{fg}/(v_f\,\mu_f)^2$
π_3= Bubble formation = $d\sigma/(v_f\,\mu_f 2)$
π_4= Inlet pressure = $d^2P_{in}\,/(v_f\,\mu_f 2)$
$\pi_{5subcooled}$= Inlet condition (subcooled) = $d^2c_{p,f}\,\Delta t_{sc}(v_f\,\mu_f)^2$
$\pi_{5quality}$=Inlet condition (quality) = x
π_6= Density effect = v_g/v_f
π_7= Viscous effect = $(\mu_f - \mu_g)/\mu_g$
π_8= Flow rate = $m/(d\mu_f)$
exp1 = -0.484 (subcooled), -0.635 (quality)
exp2 = -0.824 (subcooled), -0.189 (quality)
exp3 = 0 (subcooled), 0 (quality)
exp4 = 1.369 (subcooled), 0.645 (quality)
exp5 = 0.0187 (subcooled), -0.163 (quality)
exp6 = 0.773 (subcooled), -0.213 (quality)
exp7 = 0.265 (subcooled), -0.483 (quality)
exp8 = 1 (subcooled), 1 (quality)
f_Q = Heat loss factor
V^o_{av} = Actual induced volume of the suction vapor at suction pressure (cfm)
V^o_P = Calculated displacement of the compressor (cfm)
η = Efficiency

SUBSCRIPTS

dsh = Desuperheated
2ph = Two phase region
sub = Subcooled region
in = Entering
out = Exiting
o = Outside
I = Tube inlet

air = Air side properties
E= Evaporating
S = Electronic heat source
IN = Interface between cold plate and electronic heat source
g = Gas phase
f = Liquid phase
fg = Difference between saturated vapor and liquid
c = Caloric temperature
tot = Total quantity
a = Air
v= Volumetric
isen = Isentropic
co = Cold plate
cp = Compression
com = Compressor
mo = Motor
suc = Suction
r = Refrigerant
F = Fin

INTRODUCTION

Removal of heat has become one of the most important challenging issues facing computer system designers today. As the rate of power dissipations from electronics components continue to increase, standard conduction and forced-air convection techniques no longer provide adequate cooling for sophisticated electronic systems. The reliability of the electronic system will suffer if high temperatures are permitted to exit. Conventional thermal control schemes such as air-cooling with fans, thermoelectric cooling, heat pipes, and vapor chambers have either reached their practical application limit or are soon to become impractical for recently emerged electronic components. When standard cooling methods are no longer adequate, exotic new methods are often utilized.

The computer industry is seriously considering utilizing refrigeration as an alternative to air for cooling future high power processors. A number of attempts to incorporate refrigeration in the computer cooling in past have been reported in open literature (ref.1-6). Utilization of refrigeration as a viable cooling solution has been traditionally challenged by concerns such as system operating conditions and heat capacity, overall system efficiency, reliability, packaging, size and cost. Also, system RAS is an issue of critical importance. For an active cooling system such as refrigeration to be able to compete and replace traditionally well-understood passive air-cooled technologies, design of cooling system packaging gains prime importance. The packaging should be such that the end user should not be exposed to the refrigerant or liquid flow in the refrigeration circuitry and the entire refrigeration unit should be viewed as an air-cooled unit.

Another fundamental issue that has prevented successful widespread utilization of refrigeration cooling as an alternative in the computer industry has been lack of existence of a compressor technology that can deliver reliable and continuous performance while minimizing many design concerns that have existed with traditional mechanical vapor compression systems. These concerns include issues such as the physical size of the compressor, noise, vibration, high head dissipation, low energy efficiency, unidirectional operation and cost. A recently developed miniature compressor has promises to resolve many of the above-mentioned issues. Oil-less or low viscosity oil -lubricated operation is a requirement in refrigeration-assisted CPU cooling of computers due to orientation-independent design and utilization of the central processing units of computers.

The purpose of this work is to develop an accurate engineering model to assist engineers with an accurate design and rapid analysis of refrigeration cooling systems of computers. This paper presents an integrated analytical tool for design and rating of a CPU spot-cooling refrigeration system for computers and data processing systems. The refrigeration simulation model is based on a steady-state mathematical model for simulating the performance of a mechanical compression refrigeration system composed of a compressor, condenser, evaporator, capillary tube and associated piping system.

MATHEMATICAL MODEL

Theoretical analysis of the miniature CPU-refrigeration cooling system presented in this work relies on a steady-state vapor compression simulation model developed by Fischer (ref. 7). In this work, the original model, which was developed for a heat pump system, is modified to make it appropriate for simulating the vapor compression system of a miniature CPU-cooler. In addition, specifying a multi-zone heat exchanger model for the condenser and evaporator further expands the original model. The modifications to the original refrigeration model of Fischer (ref. 7) are explained below.

CONDENSER MODEL

Analytical modeling of the condenser is based on a three-zone heat exchanger model originally developed by Kempaik (ref. 8). The condenser is divided into three zones of desuperheated, two-phase and subcooled regions. Refrigerant enters the condenser at temperature, pressure and flow rate of T_{in}, P_{in} and $m_r^{'}$, respectively. At steady-state conditions, the refrigerant leaves the desuperheated region of the condenser at saturated vapor conditions of T_g and P_g. The portion of the condenser in this region is designated as L_{dsh}, which is also shown as $F_{dsh}L$, where F_{dsh} is the fraction of the entire condenser in the desuperheated region and L is the length of the condenser traveled by the refrigerant. Similarly, the other two regions are designated with two length factors of L_{2p} and L_{sub}. Refrigerant leaves the two-phase region at temperature and pressure of T_f and P_f, respectively. Upon further cooling of the refrigerant, it leaves the condenser at subcooled temperature and pressure values designated as T_{sub} and P_{sub}, respectively. Heat transfer in each region is given as:

$$Q_{dsh} = m_r^{'}\left(i_{in} - i_g\right) \qquad (1)$$

$$Q_{2ph} = m_r^{'}\left(i_g - i_f\right) \qquad (2)$$

$$Q_{sub} = m_r^{'}\left(i_f - i_{out}\right) \qquad (3)$$

Based on heat transfer relations for each region of the condenser regions, we have:

$$Q_{dsh} = F_{dsh} LU_{dsh}\left(T_{c,dsh} - T_{c,air}\right) \qquad (4)$$

$$Q_{2ph} = F_{2ph} LU_{2ph}\left(T_{c,2ph} - T_{c,air}\right) \qquad (5)$$

$$Q_{sub} = F_{sub} LU_{sub}\left(T_{c,sub} - T_{c,air}\right) \qquad (6)$$

Where, subscript "c" refers to the caloric temperatures of air or that of refrigerant at each region. It is then assumed that the products of heat transfer length-overall heat transfer coefficients, LU_i, is a function of the air volumetric flow rate, $V^{'}$, as:

$$\frac{1}{LU_i} = b_i + \frac{C_i}{V^{'0.5}} \qquad (7)$$

Combining the above equations along with corresponding heat transfer equations for each region, we utilize a least square method for determining the values of constants given in the above equation. Similarly, pressure drop for the refrigerant flowing along the condenser is divided into three regions of desuperheated, two-phase and sub-cooled regions. Pressure drop in each of the designated regions is given as:

$$\Delta P_{dsh} = \frac{m_r^{'2}}{A^2\rho_g} - \frac{m_r^{'2}}{A^2\rho_{in}} + \left(b + \frac{C_{dsh}}{m_r^{'0.237}}\right) L \frac{F_{dsh}}{d}\frac{m_r^{'2}}{2\rho_{in}A^2} \qquad (8)$$

$$\Delta P_{2ph} = \frac{m_r^{'2}}{A^2\rho_f} - \frac{m_r^{'2}}{A^2\rho_g} + \left(b + \frac{C_{2ph}}{m_r^{'0.237}}\right) L \frac{F_{2ph}}{d}\frac{m_r^{'2}}{2\rho_f A^2} \qquad (9)$$

$$\Delta P_{sub} = \frac{m_r^{'2}}{A^2\rho_{out}} - \frac{m_r^{'2}}{A^2\rho_f} + \left(\frac{C_{sub}}{m_r^{'0.25}}\right) L \frac{F_{sub}}{d}\frac{m_r^{'2}}{2\rho_f A^2} \qquad (10)$$

Again, the values of the constants are determined using a statistical approach such as a least square method. Total heat transfer and pressure drops are then calculated as:

$$Q_{tot} = Q_{dsh} + Q_{2ph} + Q_{sub} \qquad (11)$$

$$P_{tot} = P_{dsh} + P_{2ph} + P_{sub} \qquad (12)$$

EVAPORATOR MODEL

The evaporator is where the incoming air stream is cooled or heat is absorbed directly from the CPU by boiling of refrigerant in the cold-plate evaporator attached to the CPU. The evaporator experiences space constraints since it has to be placed tightly over the CPU. Due to space constraints for refrigeration components in a computer server, there is a need for a precise tube-by-tube method for design of the evaporator coil. Modifications are made to an existing evaporator heat exchanger model (ref. 9) which is based on a detailed geometrical simulation in which heat transfer and pressure drops resulting from flow of refrigerant through individual path and lines of flow are analytically modeled. For the refrigerant, a multi-region flow model consisting of a two-phase and an all-vapor flow model, which considers existence of annular and dispersed two-phase flow regimes is used.

Two flow regimes of annular and dispersed are distinguished by the quality of refrigerant in the evaporator cold plate. Refrigerant flow quality of up to 0.85 is considered to represent annular and refrigerant quality of between 0.85 and 1.0 is considered to represent dispersed flow regimes. Flow quality of greater than 1.0 constitutes superheated flow regime. For annular flow regime,

$$Q = m'_a C_{p,a} (t_i - T_i)\left(1 - \exp\left(-\frac{UA_o}{m'_a C_{p,a}}\right)\right). \quad (13)$$

For dispersed flow regime,

$$Q = m'_a C_{p,a} (1 - ANNUL)(t_i - T_i)\left(1 - \exp\left(-\frac{UA_o}{m'_a C_{p,a}}\right)\right) \quad (14)$$

For single phase superheated flow of refrigerant in the evaporator heat exchanger,

$$Q = m'_r C_{p,r} (t_i - T_i)$$

$$\left(1 - \exp\left(-\frac{(1 - ANNUL - XDRY) m'_a C_{p,a}}{m'_r C_{p,r}}\right)\right)\left(1 - \exp\left(-\frac{UA_o}{m'_a C_{p,a}}\right)\right) \quad (15)$$

Where, the fraction of heat exchanger in annular flow regime, ANNUL, is calculated as,

$$ANNUL = \frac{m'_r (h_{r,0.85} - h_{r,i})}{m'_a C_{p,a} (t_i - T_i)}\left(1 - \exp\left(-\frac{UA_o}{m'_a C_{p,a}}\right)\right) \quad (16)$$

The fraction of flow in dispersed regime, XDRY, is calculated as,

$$XDRY = \frac{m'_r (h_{r,v} - h_{r,i})}{m'_a C_{p,a} (1 - ANNUL)(t_i - T_i)} \cdot \left(1 - \exp\left(-\frac{UA_o}{m'_a C_{p,a}}\right)\right) \quad (17)$$

For the case of an attached cold plate heat exchanger, the formulation becomes,

$$Q = UA(T_E - T_S) \quad (18)$$

$$U.A = \frac{1}{\frac{1}{(\eta_F hA)_R} + \frac{L}{(KA)_{CO}} + \frac{1}{(hA)_{IN}}} \quad (19)$$

The overall heat transfer coefficient, U, consists of different resistance terms for the flow of heat. Heat transfer calculations employ appropriate inside tube heat transfer correlations for annular, dispersed, and single-phase superheated vapor flows available from literature (ref. 9).

COMPRESSOR MODEL

The miniature compressor is a positive displacement, piston-type compressor in which the piston is driven directly by a motor, rather than by a rotary motor coupled to a mechanical mechanism as in a conventional reciprocating compressor.

A theoretical isentropic efficiency-loss compressor mathematical model is used for simulating the performance of the compressor in this work. In its ideal form, a reciprocating compressor is assumed to operate in a reversible adiabatic manner with pressure losses in the valves and manifolds are considered negligible. In cycle analysis, losses external to the compression process like machine friction and motor winding losses are frequently incorporated into the pressure–enthalpy diagram as cycle inefficiencies relative to the isentropic process (Figure 1), implying that isentropic compression is the most efficient path between suction and delivery conditions. The lower limit of power consumption is represented on the P-h diagram as the isentropic process but this is only true where no heat transfer to the working fluid takes place.

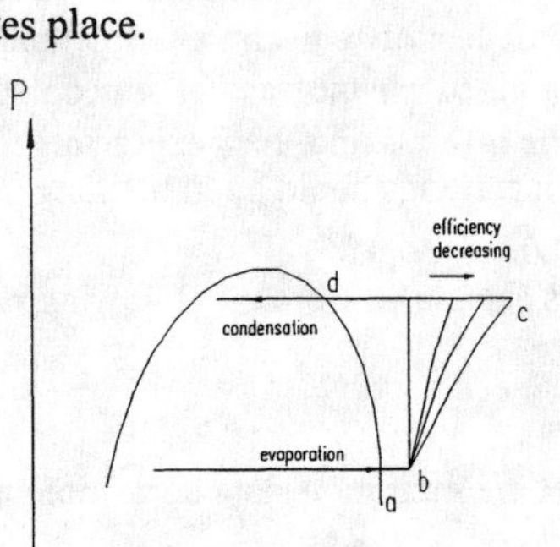

Figure 1 Work input represented by enthalpy difference (c–b) on a P–h diagram

Figure 1: p-h diagram of an actual and an ideal vapor compression

Volumetric efficiency of a compressor is defined as the ratio of the actual mass of the vapor compressed to the calculated displacement of the compressor. It describes how efficient a compressor compresses refrigerant volume for the given available displacement of the compressor, or:

$$\eta_v = \frac{V^o_{av}}{V^o_P} \quad (18)$$

Isentropic efficiency η_{isen}, compression efficiency η_{cp}, compressor efficiency η_{comp}, and mechanical efficiency η_{mec} are defined as:

$$\eta_{isen} = \frac{(i_2 - i_1)}{(i_2^{'} - i_1)} = \eta_{cp}\eta_{mec} = \eta_{comp} \quad (19)$$

$$\eta_{cp} = \frac{W_{sen}}{W_v} \quad (20)$$

$$\eta_{mec} = \frac{W_v}{W_{comp}} \quad (21)$$

where i_1, i_2, i'_2 are enthalpies of the suction vapor, ideal discharged hot gas, and actual discharged hot gas, respectively. W_{isen}, W_v, W_{com} are isentropic work, work delivered to the vapor refrigerant, and work delivered to the compressor, respectively. The actual power input to the compressor P_{comp} is given as:

$$P_{com} = \frac{m_r^o (i_2 - i_1)}{\eta_{isen}\eta_{mo}} \quad (22)$$

$$m_r^o = V_p^o \eta_v \rho_{suc} \quad (23)$$

$$\eta_{mo} = \frac{P_{com}}{P_{mo}} \quad (24)$$

Ratio of heat loss to the energy consumption by the compressor is known as the heat loss factor, f_Q.

CAPILLARY TUBE

The capillary tube has the duty of controlling the flow of the refrigerant in the refrigeration cycle. Proper design of the capillary tube can improve and optimize system performance drastically. It is a long thin tube placed between the condenser and the evaporator. The small diameter and long length of the capillary tube produces a large pressure drop. Effective control of the system results because the tube allows flow of liquid more readily than vapor. The main advantage of the capillary tube over other flow restricting and controlling devices used in refrigeration systems is its simplicity and low cost.

Subcooled liquid refrigerant enters the capillary tube and pressure losses due to friction causes the refrigerant to reach saturation at some point in the tube. At that point the pressure loss becomes much more pronounced as vapor forms and the refrigerant temperature decreases rapidly to the tube exit. The amount of refrigerant flowing through the tube depends on the overall pressure difference between the condenser and the evaporator and on the saturation pressure of the refrigerant entering the tube.

Although there are calculation procedures available for the sizing of capillary tubes, the correlation by Wolf et al. (ref. 10) is adopted here. The Buckingham π-theorem was used for finding a number of relevant π-terms, and the exponent of each term was determined from the experimental data. Different π-terms were determined for the case of subcooled conditions at the tube entrance and two-phase (quality) conditions respectively. The definitions of the π-terms and their exponents for the two cases are given in the nomenclature. For the case of subcooled conditions at the inlet, the subcooling should be in the range $1 < \Delta t_{sc} < 17$ and for the case of two-phase flow at the inlet, the vapor quality should be in the range $0.03 < x < 0.25$.

The mass flow rate of refrigerant is determined from π_8, which is calculated from the following relation:

$$\pi_8 = B.\pi_1^{\exp 1}.\pi_2^{\exp 2}.\pi_3^{\exp 3}.\pi_4^{\exp 4}.\pi_5^{\exp 5}.\pi_6^{\exp 6}.\pi_7^{\exp 7} \quad (22)$$

All fluid properties, both liquid and vapor, should be evaluated at the saturation state corresponding to the fluid temperature at the tube inlet.

METHOD OF SOLUTION

Steady state simulation of the refrigeration system is performed by simultaneously solving the system of equations representing the component in accordance to the flow chart shown in Fig. 2. The flow chart in Figure 2 indicates that discharge pressure (P_{dis}), suction pressure (P_{suc}) and the degree of superheat (Δt_{sup}) are corrected based on mass flow rate, enthalpy energy and system total charge mass balance equations, respectively. The entire system of equations is formulated and solved using EES (ref. 11) which includes NIST's refrigeration property routine REFPROP (ref. 12), enabling us to perform system simulations using variety of refrigerants.

RESULTS AND DISCUSSIONS

Figure 3 shows the results of simulation analysis for a CPU spot-cooling refrigeration system using a miniature compressor, a compact air-cooled condenser, a cold plate evaporator and capillary tube. The working refrigerant is assumed to be R-134a and the calculations are performed for an assumed 86 °C junction temperature. The Figure shows a schematic of the result of cycle analysis on an enthalpy-pressure diagram. The calculated COP of the system with the specified configurations is 3.0.

Figure 4 shows the effect of varying system refrigerant from R-134a to a number of commercially available refrigeration and air-conditioning refrigerants. As can be seen, the highest system performance is obtained for the case of Ammonia, with R-22 (a favorable air-conditioning refrigerant) giving slightly less overall system COPs than R-134a. Considering various considerations such as thermodynamics, physical, and chemical characteristics, safety issues, and the effect of the refrigerant on the environment (ozone depletion and global warming potentials), as well as overall system size and cost issues, R-134a is considered a favorite refrigerant for CPU-cooling using vapor compression refrigeration systems.

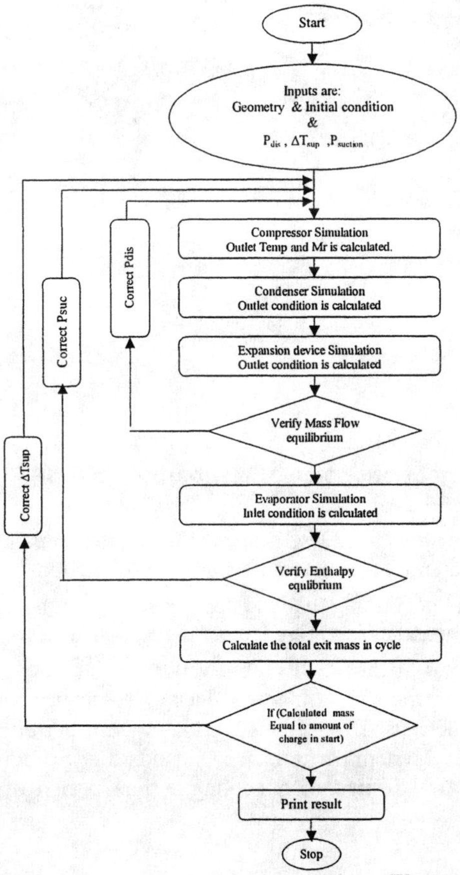

Figure 2: Computer simulation flow chart

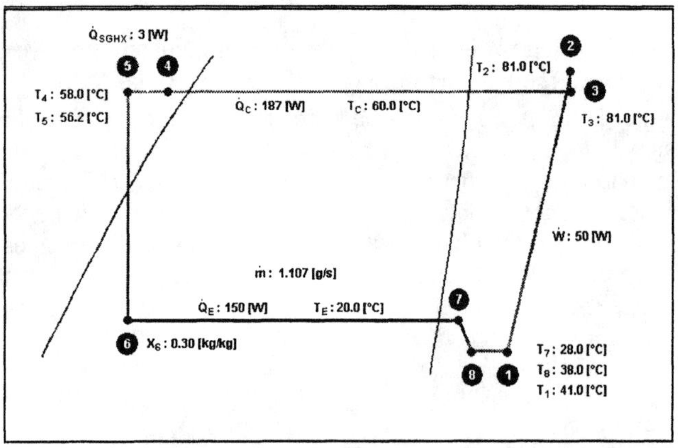

Figure 3: Cycle analysis of a typical CPU spot-cooler

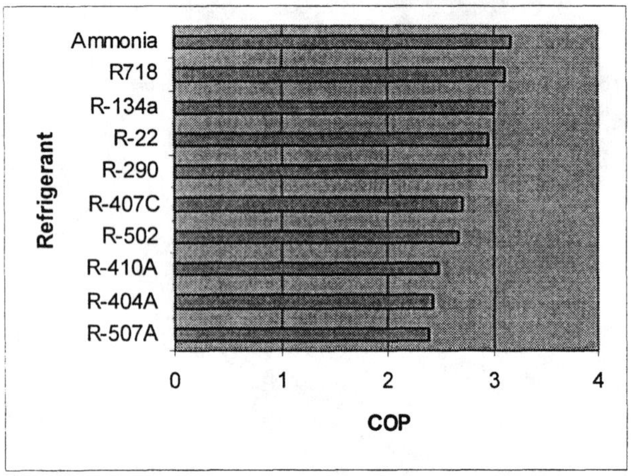

Figure 4: Effect of choice of refrigerant on COP

Figures 5 and 6 show the effect of varying the evaporating temperature from the specified 20 °C and its effect on the overall system COP and the condenser load, Q_{cond}. It can be seen that increasing the evaporating temperature for a fixed junction temperature and evaporator load, causes an increase in the system efficiency, while decreasing the load and consequently the size of the condenser. On the other hand, Figure 7 shows that for a fixed Q_{evap} and junction and evaporating temperatures, condenser temperature has much greater effect on the overall system efficiency, with much higher COPs observed at lower condensing temperatures. Lower condenser temperatures can be obtained at lower ambient temperatures, at higher values of volumetric airflow rates caused by system fans and with more efficient condenser heat exchanger designs. Even though, actual system performance parameters shall be evaluated using laboratory test equipment based on evaluation of system performance at standard test conditions, the results of this work help system designers obtain greater insights into practical refrigeration cooling system design of next generation of high heat dissipating CPUs.

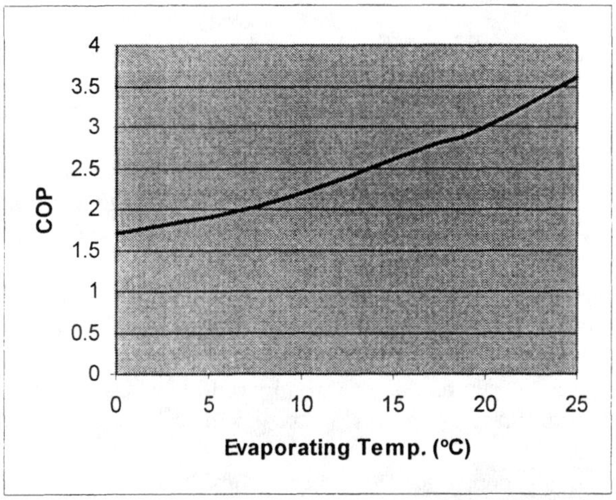

Figure 5: Effect of variation of T_{evap} on COP

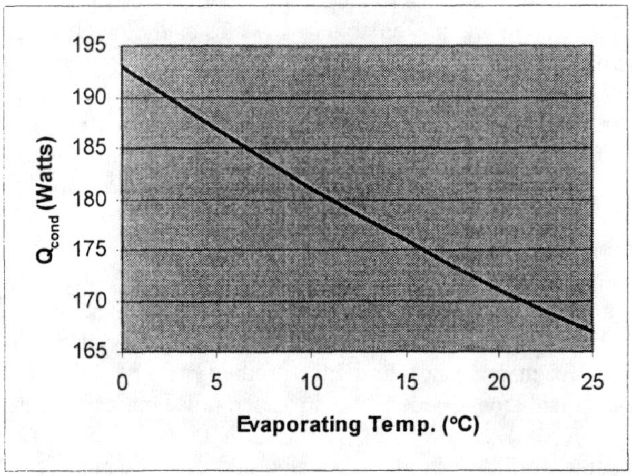

Figure 6: Effect of variation of Tevap on Qcond

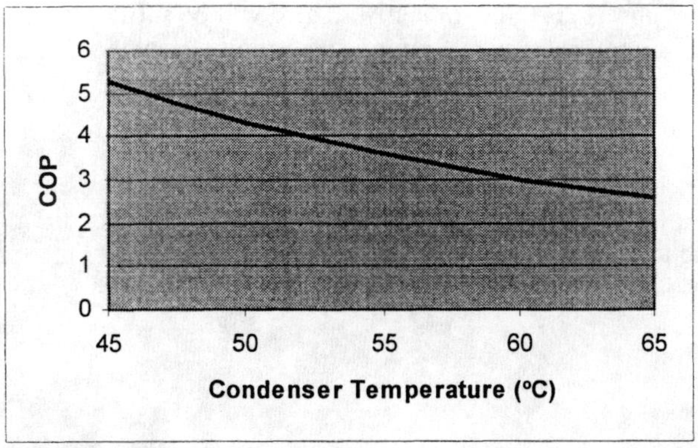

Figure 7: Effect of variation of T_{cond} on COP

RELIABILITY

Public pressure for environmental sensitivity and the current energy crisis in California forces the industry to consider energy efficiency along with performance, reliability, maintainability and cost effectiveness when designing and purchasing a cooling system for their computers. The human side of using refrigeration for cooling the next generation CPUs is of prime importance.

Reliability is the probability that an item will perform satisfactorily for a specified period of time under specified operating conditions. The compressor is the least reliable of the components of the vapor-compression cycle so this section will limit its discussion to compressor reliability issues. The condenser and the evaporator are important but are generally more reliable components. Reliability data available for compressors indicates that the mean-time-between-failures is roughly 77,000 hours for commercial equipment at atmospheric conditions. The life cycle of compressors relates to the wear-out of life limited components and can vary from 80,000 to 240,000 hours. Because the design is being optimized to replace some life limiting components, the L_{10} for this compressor has been estimated to be 100,000 hours at 50 °C ambient. (L_{10} means the lifetime at which 10% of the mechanical devices will have failed.) This value will be validated during detailed design. Generally there are very low failures during the compressor useful life; however, there is fast wear out at the end of life due to fatigue and corrosion. This is acceptable for the computer industry where the machines are obsolete at the end of life.

The most important thing for the computer industry is to eliminate isolated, random events as well as the initial failures due to manufacturing faults. Performance of compressors is tied to the details of manufacturing control. There is a limit to the extent and expense that commercial manufacturers can undertake and still make a profit. The developers must work closely with the manufacturer to assure the required reliability while holding down costs. Some examples of manufacturing induced defects include trapped moisture (corrosion), cracks/voids, residual cleaning chemicals, dust, thermal stress and bonding. The required reliability is equal to the useful life of the computer. In other words, cooler failures must not contribute to system unavailability.

Highest reliability should be designed into individual components such as pistons, cylinder, gears, seals, rings, bearings, springs, valves and motor. This means designing to minimize the deterioration wear out, fatigue wear out, corrosion, leakage, cracks, pressure drop across valves and other potential causes of failures during the operational life of the compressor. The use of gas bearings assures a frictionless contact between piston and cylinder. This reduces friction losses as well as noise. An opposed piston configuration cancels the vibration forces because the moving masses are balanced. The mechanical springs are designed to meet actual life requirements eliminating their material fatigue limits and lowering stress levels.

There are other improvements in design which will increase the reliability. The compressor must be rated for higher temperatures for longer, more reliable periods between maintenance. The motor insulation must be rated at a higher temperature than operational temperature. Compressor valves are a top cause of failure of refrigeration cycles. Designing compressor valves for high reliability is difficult due to overlapping factors that effect performance. Of course there are unavoidable losses such as the fact that valves do not open and close instantaneously, that the moving valve elements do not stay fully open and the inherent horsepower losses. When valves fail they either seize, fail to open because the spring has broken or they simply disintegrate. Flapper or Poppet valves are considered viable options for design of the miniature compressor. Flapper valves contain a bi-metal thermo-spring that allows the valve to open as the exhaust warms. These valves have a high failure rate and are being investigated in detail for their viability.

SERVICEABILITY

Leakage of refrigerant is a major concern for computer hardware, and the environment. A serviceability model will be required to eliminate the possibility of mistakes such as a service personnel putting a screwdriver through a refrigerant line.

The refrigeration system would be as invisible to the user as an air cooled thermal management system. Utilization of a miniature refrigeration system requires no external plumbing lines since the refrigeration system is a self contained system. An oil-less compressor needs less maintenance. The goal is to have no preventative maintenance during the product life.

Noise levels are reduced by the use of gas bearings. Mechanical isolation to reduce vibration and noise must be balanced against the need to reject the heat of compression.

Other serviceability issues are the need for redundancy in modules, fail over in the data center and dynamic recovery. The use of state-of-the-art electronic controls and sensing will provide the required data for dynamic recovery. The controls and sensing data should provide sufficient information to determine if the refrigeration unit is operating properly. There will be sufficient warning of failure modes to reconfigure the unit out of the running domain, either by soft shutdown or moving the activity to another processor.

CHOICE OF REFRIGERANT

R-134a is the refrigerant of choice since it is widely available, is environmentally friendly and has operating pressures which allow for smaller swept volumes and therefore decreases the size of the compressor. The system must be rated for the required refrigerant burst pressure, the compressor displacement must be reduced for a given motor horsepower and the shell burst pressure must be rated for the refrigerants higher pressure. A failed or obsolete refrigeration unit should have all refrigerant removed and disposed of properly in accordance with the law.

LIFE-CYCLE ANALYSIS ISSUES

There is no statistical or field data to base a reliability or life cycle cost analysis on. The life cycle costs rely on reliability and maintainability issues to convert engineering technology into a monetary value. Procurement cost represents a small fraction of the total cost during the life of the compressor if reliability and maintainability are not sustained throughout its useful life. Accelerated testing, assembly and servicing of the refrigeration unit will cost more than a conventional air-cooled system.

CONCLUSIONS

This paper presents the result of a study conducted for simulating the performance for design and parametric sensitivity studies of a refrigeration CPU-cooling system. The analysis takes advantage of newly developed piston-free linear vapor compression compressor, which curtails many issues that existed with utilization of such systems for computer cooling applications.

Analysis of the refrigeration cycle is based on a steady-state modeling of four distinct components of the cooling system. Multi-zone heat exchanger mathematical models for simulating the condenser and evaporator heat exchangers, a physically-based efficiency model for predicting the performance of the linear compressor and empirical correlations for modeling the flow rate through the capillary tube are put together in a efficient iterative computer code.

Using the simulation program enables us to study effects of design parameters such as the choice of working refrigerant, evaporating and condensing temperatures on system components and overall efficiency.

REFERENCES

1. Ashraf, N.S., Carter, H.C., Casey, K., Chow, L.C., Corban, S., Drost, M.K., Gumm, A.J., Hao, Z., Hasan, A.Q, Kapat, J.S., Krammer, L., Newton, M., Sundaram, K.B., Kapat, J.S., Vaidya, J., Yerkes, K., Wong, C.C., "Design and Analysis of a Meso-Scale Refrigerator," ASME International Mechanical Engineering Congress & Exposition, Nashville, Tennessee, November, 1999.
2. Shannon, M.A., Philpott, M.L., Miller, N.R., Bullard, C.W., Beebe, D.J., Jacobi, A.M., Hrnjak, P.S., Saif, T., Aluru, N., Sehitoglu, H., Rockett, A., Economy, J., "Integrated Mesocopic Cooler Circuits (IMCCS)," Proceedings of the ASME Advanced Energy Systems Division, AES-Vol. 39, pp. 75-82, 1999.
3. Carter, H.C., Chow, L.C., Kapat, J.S., Laveau, A., Sundaram K.B., Vaidya, J. "Component fabrication and testing for a meso-scale refrigerator," AIAA Paper No. 99-4514, 1999.
4. Laveau, A., Kapat, J.S. Chow, L.C., Enikov, E., Sundaram K.B., "Design, analysis and fabrication of a meso-scale centrifugal compressor," ASME International Mechanical Engineering Congress and Exposition, Orlando, Florida, November 5-10, 2000.
5. Patrick E. Phelan, Victor Chiriac, Tien-Yu Tom Lee, "Current and future miniature refrigeration cooling technologies for high power microelectronics", 17th IEEE SEMI-THERM Symposium, 2001.
6. Trujillo, R., Mou, J.I., Phelan, P.E., Chua, D.S., "A prototype mesoscale compressor using electrostrictive actuation," to be submitted to Energy Conservation and Management, 2001.
7. Fischer, S. K., and Rice, C.K., " The Oak Ridge Heat Pump Models: 1. A Steady-State Computer Design Model of Air-to-Air Heat Pumps", ORNL/COM-90/R1, Oak Ridge National Laboratory, Oak Ridge, Tennessee, 1983.
8. Kempiak, M.J., Crawford, R. R., "Three-zone, steady-state modeling of a mobile air-conditioning condenser", ASHRAE Transactions, Volume 98, Part 1, pp. 475-488, 1992.
9. Domanski, A.P., "Simulation of an evaporator with nonuniform one-dimensional air distribution", *ASHRAE Transactions*, Vol. 97, part 1: pp. 793-802, 1991.
10. Wolf, D. A., Bittle, R. R., Pate, M. B., " Generalized Performance prediction method for adiabatic capillary tubes," HVAC&R Research, Volume 4, Issue 1, pp. 27-43, 1998.
11. Klien S.A. Alvarado F.L., "EES-engineering equation solver, 4406 Fox Bluff Rd., Middleton, WI 53562, F-Chart Software, 1998.
12. Gallagher, J., Huber M., Morrison G., McLinden M., NIST standard refernce database 23, NIST thermodynamic properties of refrigerants and refrigerant mixtures database (REFPROP), National Institute of Standards and Technology, 1993.
13.

Proceedings of
2001 ASME International Mechanical Engineering Congress and Exposition
November 11–16, 2001, New York, NY
EPP-Vol. 1

IMECE2001/EPP-24711

A Mixed Thermal Convection Criteria for a Printed Circuit Board

E. Kehoe, R. Grimes, M. Davies,
Stokes Research Institute,
Department of Mechanical and Aeronautical Engineering,
University of Limerick, Ireland
Phone: +353 61 202471 Fax: +353 61 202393
E-mail: emily.kehoe@ul.ie

ABSTRACT

Mixed convection is a fundamentally significant heat transfer problem that occurs in a wide range of industrial and technological applications. For example, it is of great interest to thermal designers in the field of electronic cooling. Given that its occurrence is frequent, there is a necessity to understand its physics. This investigation is focused on a horizontal cylinder in cross-flow, at low Reynolds numbers of order $10^0 - 10^2$ and high Grashof numbers of order 10^4 and at the Prandtl number for air. Particle image velocimetry and interferometric measurements for these cases clearly show the range of Gr/Re^2 for which mixed convection occurs. Infra red temperature measurements from a printed circuit board mounted in a test electronic system at similar values of Grashof and Reynolds number are then presented which exhibit similar mixed convection phenomena.

The novelty lies in using the detailed measurements from the cylinder experiments to explain the temperature distribution across the printed circuit board. To do this requires finding the correct and characteristic lengths in the similarity condition. This should therefore allow the designers of electronic systems to estimate the values of Reynolds and Grashof number for which buoyancy effects need to be considered. This in turn will enable the designer to better predict the circuit board temperature distribution and component reliability.

INTRODUCTION

Mixed convection occurs when the velocities induced by an external force, such as a fan, are of the same order as those induced by the buoyancy force. This situation most commonly occurs when a fluid such as air is forced to flow over a heated object at low velocities. Several investigations have shown that the criteria defining the boundaries of mixed convection is based upon the value of the ratio Gr/Re^2 [1, 2]. The transition occurs in the region of $Gr/Re^2=1$. There is no agreement on the values at the beginning and end of this transition. The transition is smooth, in terms of the mean Nusselt number, from free to forced flow. It follows that at transition either a d or free convection correlation can be used. It is generally accepted that for forced convection Nu α $Re^{1/2}$ and for free convection Nu α $Gr^{1/4}$. At the transition therefore $Re^{1/2}$ α $Gr^{1/4}$,quadrupling gives the determining ratio Gr/Re^2 for transition. Kehoe et al [3] have shown this in some detail for a horizontal cylinder.

This paper presents results from two independently performed investigations. Firstly, results of a fundamental investigation into mixed convection flow over a horizontal cylinder are presented. This data is then used to explain the temperature distribution on a PCB mounted in a test electronic enclosure. It is shown that at forced velocities, that are typical if those found in electronic systems, buoyancy can have a significant effect on the component temperatures.

NOMENCLATURE

Symbol	Description	Units
K	Gladestone-Dale constant	m^3kg^{-1}
L	Characteristic Length	m
N	Fringe number	-
R	Universal gas constant	$Jkg^{-1}K^{-1}$
T	Temperature	°C /K
d	Diameter	m
g	gravitational acceleration	ms^{-2}
h	Heat transfer coefficient	$Wm^{-2}K$
k	Thermal conductivity	$Wm^{-1}K^{-1}$
n	Refractive index	-
p	Pressure	Nm^{-2}
$u_?$	Forced flow approach velocity	ms^{-1}
β	Volume coefficient of expansion ($1/T_?$)	K^{-1}
?	Emissivity	-
Δ?	Path Length	m
?	Wavelength	m
?	Density	kgm^{-3}
μ	Dynamic viscosity	$kgm^{-1}s^{-1}$
Subscript		
Cyl	Cylinder	-
O	Reference condition	-
?	Quiescent / free stream conditions	**-Dimensionless Numbers**
$\overline{Nu}$	Mean Nusselt number	$\overline{h}d/k$
Nu	Local Nusselt number	hd/k
Re	Reynolds number	$u_?d?/\mu$
Gr	Grashof Number	$?^2g\beta d^3(T_{cyl}-T_?)/\mu^2$
Pr	Prandtl Number	$\mu c_p/k$

EXPERIMENTAL MODELS

Enclosure Mounted Horizontal Cylinder

The test geometry used is shown in figures 1 and 2. It consists of a cylinder mounted horizontally in a glass walled enclosure. The flow was induced by an electronic cooling fan operating at a fixed speed.A range of Reynolds numbers of 33-194 were obtained by systematically adding screens to the inlet to block the flow. Corresponding velocities varied from 0.021m/s to 0.12m/s. Dalton and Davies [4] demonstrated the difficulties involved in measuring such low velocity flows using hot wire anemometry. So a further novelty of this work is in demonstrating the use of PIV under these conditions.

The forced flow is blowing across the buoyancy induced flow. The cylinder was long enough to obtain a two dimensional flow field and small enough in geometry to not be affected by the walls of the working section. An inlet layer of honeycomb was used to straighten the flow. One 600W N&P cartridge heater, of 9.5mm and 150mm in length powered the cylinder to give a fixed Grashof number. Temperatures were measured using three calibrated T-Type (± 0.1°C) thermocouples; one mounted inside the cylinder, a second shielded thermocouple was used to measure a reference fringe temperature, and a third employed to read the ambient temperature. The power input to the heater was measured with a meter to an accuracy of 0.01V and 0.01A.

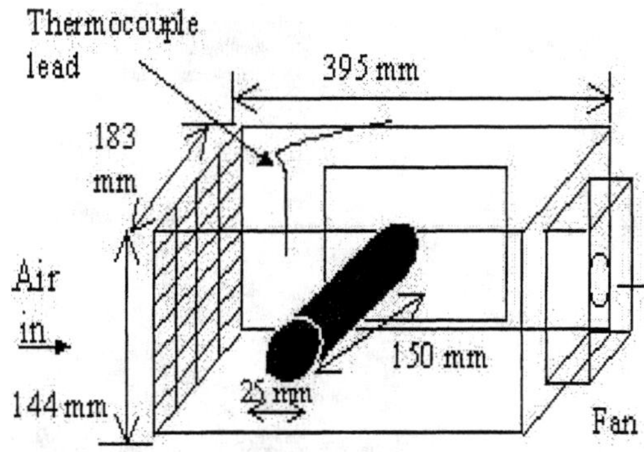

Figure 1. Front view of test enclosure containing horizontal cylinder

Figure 2. Elevation view of test enclosure containing horizontal cylinder

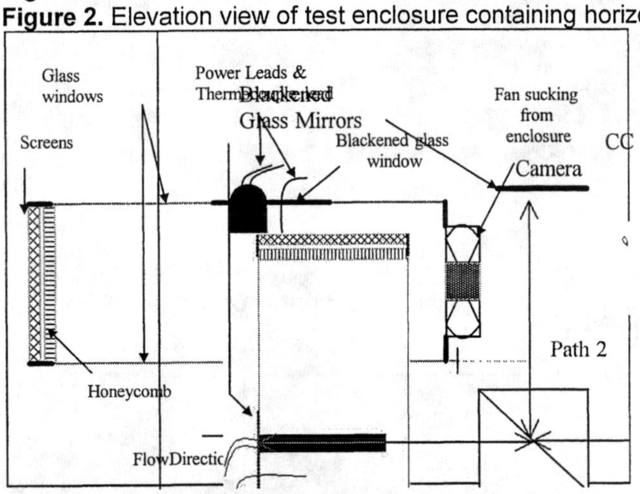

Enclosure mounted PCB

The test electronic enclosure in which PCB measurements were performed is shown in figure 3. A 5x3 array of components is mounted on the PCB. Each component comprised of a thin film heater attached to the underside of an aluminum block. Each component dissipates 3W. The dimensions of the enclosure in which the PCB is mounted are identical to those of the enclosure in which the horizontal cylinder was mounted. A fan, mounted at the outlet of the enclosure, drew air through the enclosure. The enclosure was mounted at the outlet of a fan performance test rig, which was used to control and measure the enclosure flow rate. This test rig is described in greater detail by Grimes et al [5]. Flow rate through the enclosure was $0.0236 m^3/s$ in this investigation.

To enable infra red measurements of PCB surface temperature, the wall of the enclosure on the component side of the PCB was made of polypropylene film. This film is 40% transmissive to radiation in the 8 to 12 µm wave band to which the IR camera used is sensitive. Knowing this the IR camera is calibrated to measure temperature. Full details of the test electronic enclosure and test PCB are given by Grimes and Davies [6].

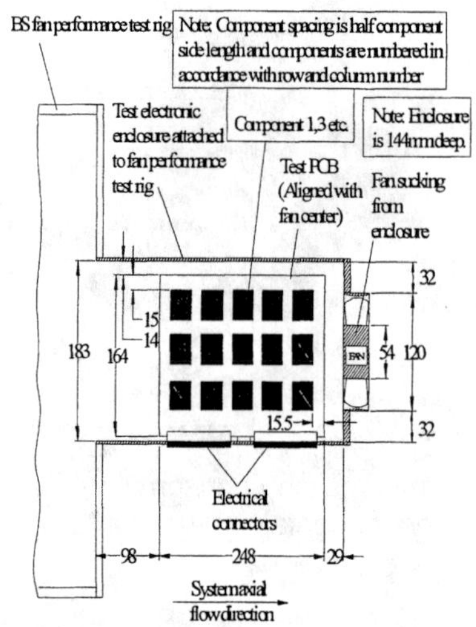

Figure 3. Test enclosure containing populated Printed Circuit Board, attached to the fan performance test rig. Dimensions are in mm. The mean flow direction is from left to right. Component height is 2mm.

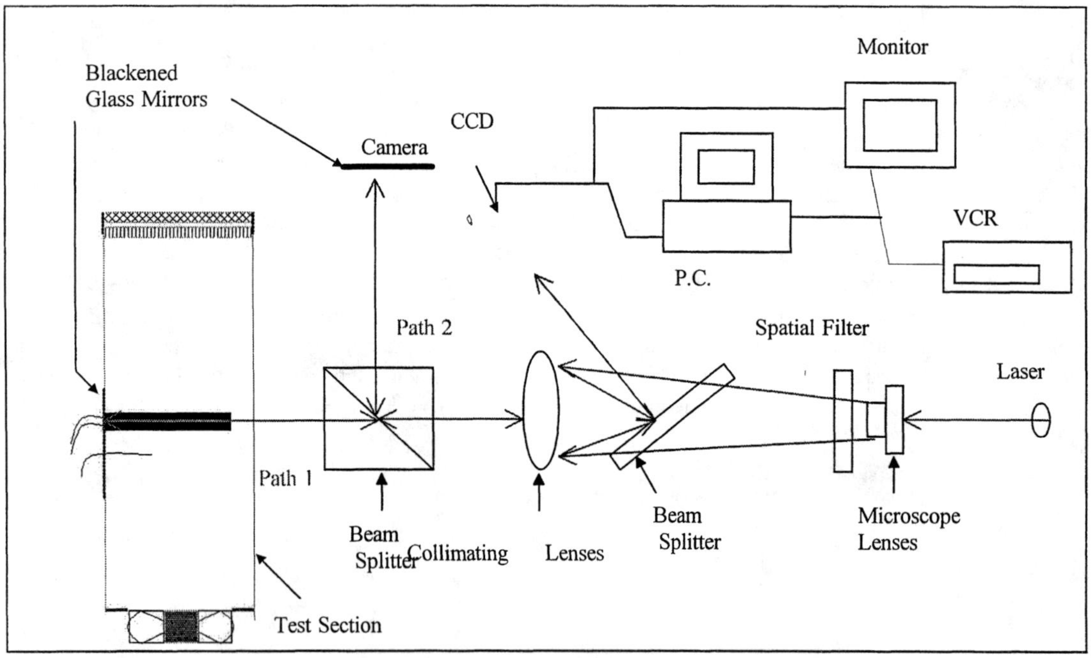

Figure 4. A Schematic of the Michelson Interferometer

MEASUREMENT TECHNIQUES

Interferometry

A Schematic of the Michelson interferometer used to measure the temperature of the air in the vicinity of the cylinder is illustrated in Figure 4. Fundamentally consisting of a laser source, beam splitter, which divides the beam into two equal intensity beams and two mirrors which return the beams to the beam splitter where they are recombined. The system is a double pass system. The interferometric measuring technique, Digital Moiré Subtraction (DMS), developed by Forno et al [7, 8], was used in conjunction with the above configuration to obtain density fringes. With this method, a pattern of fringes, called a Moiré pattern, is introduced into the initial unheated image. The initial Moiré pattern is captured, and then subtracted in real time from all subsequent interferogrames. Because the initial and all subsequent interferogrames contain the wave front distortion due to inhomogenities present in the optics, the subtraction process eliminates the additional fringes caused by these. The resultis a live interferograme containing fringes which are caused solely by the heating process. For a 2D-system [9] the change in path length is related to the refractive index of the phase object via;

$$\Delta ? = \int_{o} L(n-no)dz = N? \qquad (1)$$

For the case where the change in refractive index is due to a change in density of the phase object, as in free convection, the Gladstone-Dale equation,

$$n-no = K(?- ?o) \qquad (2)$$

relates the two. Integrating (1) and combining with (2) gives

$$?_{N} - ?o = N?/KL \qquad (3)$$

The perfect gas law

$$p = ?oRT \qquad (4)$$

is used then to determine the fringe temperature assuming that changes in density due to pressure are negligible. Full details of the accuracy of the measurements are given by Kehoe et al [3].

Particle Image Velocimetry

The velocity flow field was measured using Particle Image Velocimetry (PIV), this gives details of the full velocity field about the horizontal cylinder and PCB. PIV essentially measures the displacement of particles in a fluid over a known time interval, and hence determines the velocity. This is achieved by seeding the fluid flow with suitable particles, and obtaining two photographs of the fluid over a known time interval. To obtain the photographs, the flow is illuminated with a sheet of laser light from an Nd -YAG pulsed laser (10ns). The laser light scatters off the seeding particles, which allows for the particles to be readily detected as bright spots against a dark background by a CCD camera. An evaluation technique is employed to determine the displacement of a particle over the time interval between the two pulses of the laser, and hence the particle velocity is obtained. A cross correlation was employed in the present work. This essentially involves dividing each image into sub-regions called interrogation areas, which contain a minimum of five seeding particle. PIV statistically correlates the spatial shift of the particles in each interrogation area. Detailed information on the PIV setup is given by Kehoe et al [3].

Infra Red Thermography

Measurements of component and PCB surface temperature were taken using a non-intrusive infra red thermal imaging system. To capture the temperature across the surface of the entire PCB, the IR camera was placed some distance away from the system. The material in the system wall between the IR camera and the PCB had to have a uniform, high transmissivity for the component and PCB surface temperatures to be measured. A polypropylene film, met this requirement. As the transmissivity of this film was only approximately 40%, measurements of PCB surface temperature, recorded through the film were calibrated against thermocouple measurements. Full details of this calibration and the accuracy of the measurements are given by Grimes and Davies [6].

RESULTS AND DISCUSSION

Enclosure mounted horizontal cylinder

Figure 5 shows that in free convection the plume is vertical with clearly defined temperature fringes. As the Reynolds number increases the plume gradually leans over to finally form a wake at the highest tested Reynolds numbers. The mean Nusselt number can therefore be related to the change in the governing phenomena; the movement in stagnation points, the thickness of the boundary layers and the position of the points of separation. In figure 5 all of the interferograms except the highest Reynolds number cases are free convection. This is based upon the Gr/Re2=1 criteria derived for the area averaged Nusselt number and taking into consideration acceleration around the cylinder. In calculation of Gr/Re^2, cylinder diameter was used as the characteristic length and the velocity used was the local maximum cylinder value, twice the approach.

Figure 6 shows, using PIV measured streamlines, the trajectory of the inclined plume/wake at a Gr/Re^2 value similar to that experienced by the board mounted components. The area below the cylinder is its shadow cast by the laser. An inclined angle of order 10° is observed.

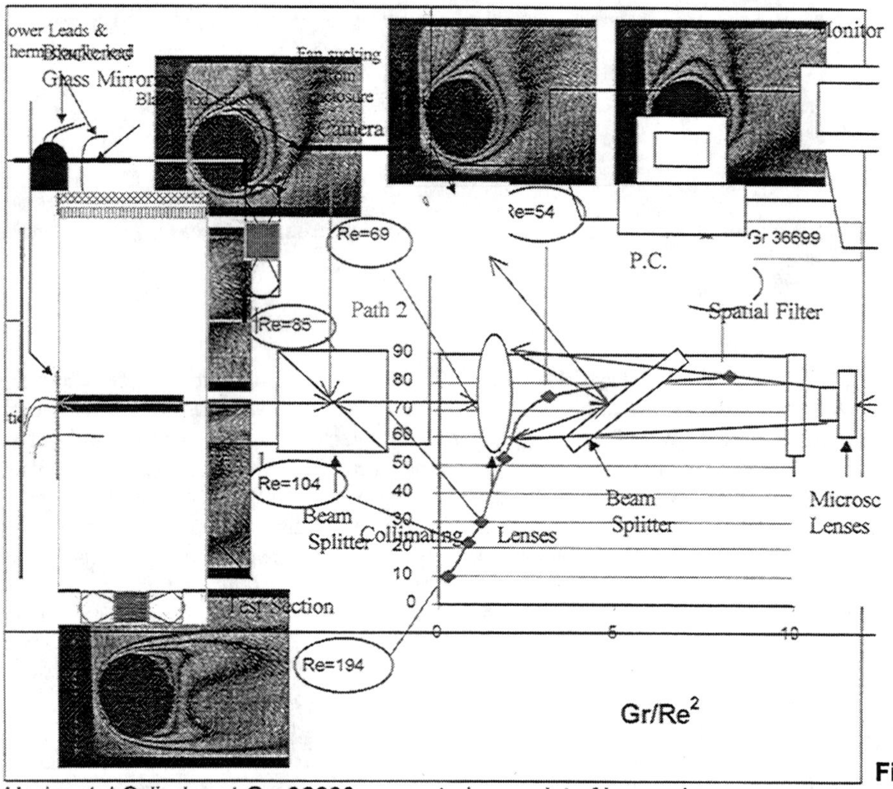

Figure 5.DMS Temperature Fringes for a Horizontal Cylinder at Gr=36699 presented on a plot of increasing Gr/Re^2 with increasing angle. $Gr/Re^2=1$ occurs a Re=104.

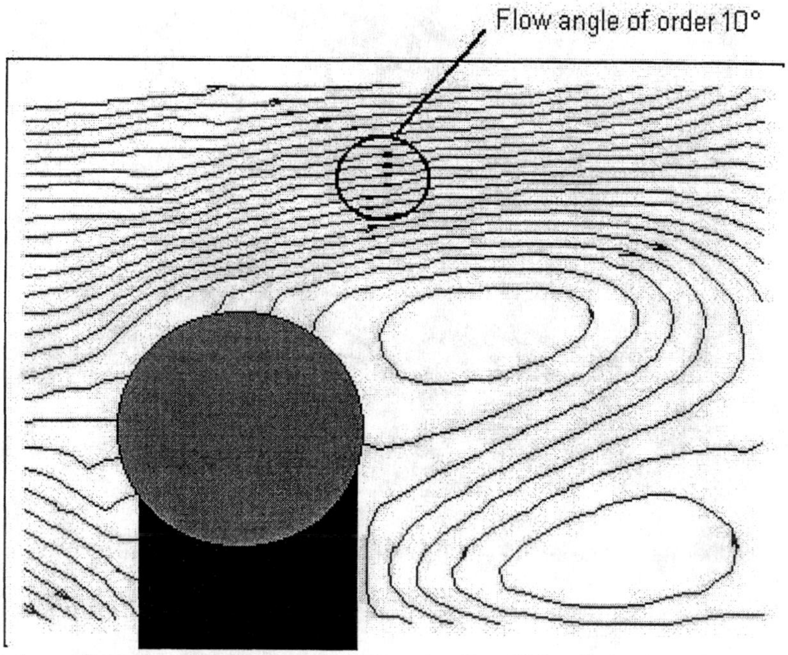

Figure 6. PIV streamlines for a Horizontal Cylinder at a Re=194, with a plume/wake angle of ≅ 10°.

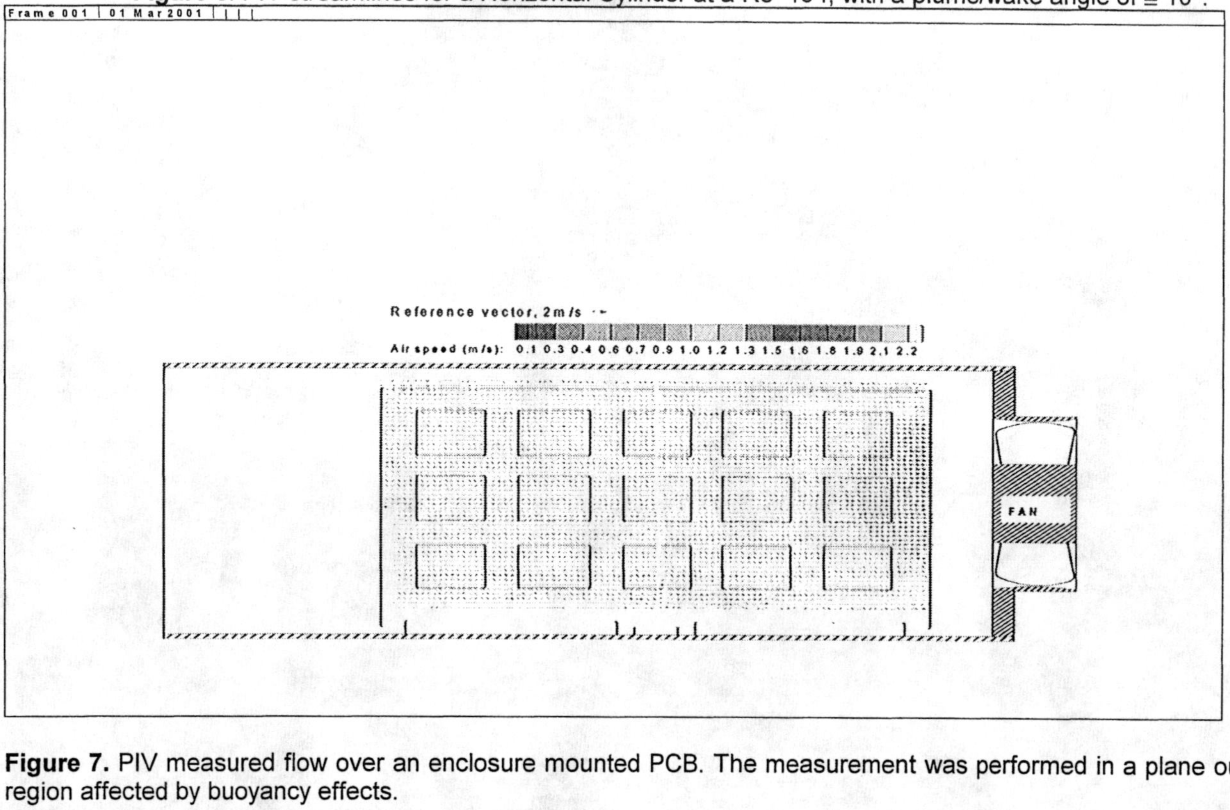

Figure 7. PIV measured flow over an enclosure mounted PCB. The measurement was performed in a plane outside the region affected by buoyancy effects.

Figure 8. Calibrated infa red thermal image of the enclosure mounted PCB.

Enclosure mounted PCB

Figure 7 shows the distribution of free stream velocity in a plane 12mm from the component side of the PCB. The flow here is clearly uniform in magnitude and direction. Free stream velocity is approximately 1.2m/s. It has often been assumed that at this velocity buoyancy effects are negligible. Figure 8 shows the temperature distribution on the enclosure mounted PCB. In uniform forced convection flow, the temperature distribution can be explained by boundary layer growth, increases in adiabatic wall temperature in the flow direction, turbulence intensity distribution and board conduction. In the configuration investigated here, in forced convection these effects should lead to a temperature distribution, which is symmetric about the center row of components (Grimes and Davies [6]). Figure 7 shows however that this temperature symmetry does not exist, temperatures of components in the bottom row being up to 12°C lower than temperatures of the corresponding upper row components. The Gr/Re^2 value for the PCB in this flow was calculated using PCB length as the characteristic length and the approach velocity of 1.2 m/s is the characteristic velocity. PCB temperature was taken to be 75°C, the average PCB component temperature. Gr/Re^2 was found to be 0.28 using these values. As shown in figure 5, convection from the cylinder was mixed at this Gr/Re^2 value. Therefore, the high temperature of the top row of components on the PCB is a result of buoyancy effects. PIV measurements of flow near the PCB indicated a vertical velocity component of up to 0.1m/s. This resulted in flow in certain areas being at approximately 10° to the axial flow direction. This flow angle is similar to that observed for the cylinder at Re=194 (see figure 6). It is apparent therefore that the angle of flow in electronic systems can be approximated by calculating the value of Gr/Re^2 and finding the flow angle downstream of the cylinder at a similar value of Gr/Re^2. This analysis can then be used to check for a buoyancy induced interaction.

CONCLUSIONS

- The angle a plume of a cylinder forms with the forced flow direction was shown to vary from 90°C to 0° as convection goes from natural to forced.
- The angle of flow over a PCB in mixed convection was shown to be similar to the angle of flow over a cylinder at a corresponding value of Gr/Re^2.
- In estimation of PCB temperature distribution, thermal designers should calculate the PCB Gr/Re^2 value and find the cylinder plume angle at the corresponding Gr/Re^2 value.

ACKNOWLEDGMENTS

The authors gratefully acknowledge PEI Technologies, the Joint Research Centre of the European Commission and the technical staff of the University of Limerick.

REFERENCES

[1] Oosthuizen, P. H. and Madan, S., "The Effect of Flow Direction on Combined Convective heat Transfer from Cylinder to Air." ASME Journal of Heat Transfer, Vol. 93, pp 240-242, 1971.

[2] Incopera, F. P., 'Fundamental of Heat and Mass Transfer.' 3d ed., John Wiley & Sons, 1990.

[3] Kehoe, E., Davies, M., Newport, D., Dalton, T., Whelan, M. and Forno, C., "Measurements of Mixed Convection Heat Transfer about a Horizontal Cylinder.", Proceedings of the 35th National Heat Transfer Conference, paper NHTC01-1417, Anaheim, California, USA, June 10-12, 2001.

[4] Dalton T., and Davies, M. 'An Experimental and Numerical Investigation of Natural Convection Plumes above a Horizontal Cylinder in an Isothermal Cubical Enclosure.' Proceedings of the 31st National Heat Transfer Conference, HTD VOL 329, Huston, Texas, USA, Vol. 7, 1996, pp 3-11.

[5] Grimes, R., Davies, M., Punch, J., Dalton, T., and Cole, R., "Modelling electronic cooling axial fan flows", *ASME Journal of Electronic Packaging,* June, 2001.

[6] Grimes, R. and Davies, M., "Aerodynamic and Thermal Investigation Into Axial Flow Fan Cooling of Electronic Systems Part I: Measurement Techniques and Steady Flow Measurements", Paper NHTC01-14110, Proceedings of the 35th National Heat Transfer Conference, Anaheim, USA, June10-12, 2001.

[7] Forno, C. and Whelan, M., 'Digital moiré subtraction for eliminating aberrations in optical and engineering interferometry'. Proceedings of. Interferometry 99, SPIE Vol. 3745, Warsaw, October 1999.

[8] Forno, C. and Whelan, M., Patent Application no. PCT/GB00/00236 (2000)

[9] Newport, D.T., 'Optical Field Measurements of Enclosure Free Convection.' PhD Thesis, University of Limerick, Ireland, 2000.

Proceedings of
2001 ASME International Mechanical Engineering Congress and Exposition
November 11–16, 2001, New York, NY
EPP-Vol. 1

IMECE2001/EPP-24712

Thermal Analysis of EPAS Power Module

Ram Farhi
Mechanical Engineer, International Rectifier, El Segundo, CA, 90245

ABSTRACT

The continuous improvement of robust design for reliability is possible by the implementation of Finite Element Methods (FEM). Thermal response of a chip-and-wire power module was studied to verify and optimize the design. The power module drives a three-phase Brushless DC motor in an Electric Power Assisted Steering (EPAS) system. It has a power output of ~1.5kW with a peak current of 125A and with thermal efficiency of 86%. The three-phase inverter is built with 12 dies (2 parallel FETs per switch), and it is constructed on an Insulated Metal Substrate (IMS) and copper lead frame.

The FEA model was verified experimentally using a Forward Looking Infra Red (FLIR) system and a threshold voltage measurement for the junction temperatures. Experimental and FEA results show the peak junction temperature is less than 130^{0}C and the thermal resistance is 1.2$^C/_W$. In addition to design changes and refinement of process prior to fabrication, the FEA approach enabled the study of competing design approaches without sacrificing valuable time.

INTRODUCTION

As the automotive industry moves to replace mechanical and hydraulic systems with electrical systems, a dramatic proliferation in the use of power electronics in the hostile under the hood environment has occurred. In this environment, high reliability, thermal efficiency and reduced size are key requirements. All of these, coupled with the trend of the constantly increasing demand for higher electric power, places stringent requirements on the thermal efficiency of automotive systems.

One such application that must meet these requirements is the Electric Power Assisted Steering (EPAS) system. EPAS system provides steering assistance to the driver on demand, using an electronically controlled three-phase brushless DC motor with torque output of 15-75Nm. This electrical system replaces the entire hydraulic power steering system. According to market studies, 13 million systems will be installed by 2004 with 28% cumulative annual growth rate (CAGR) [1]

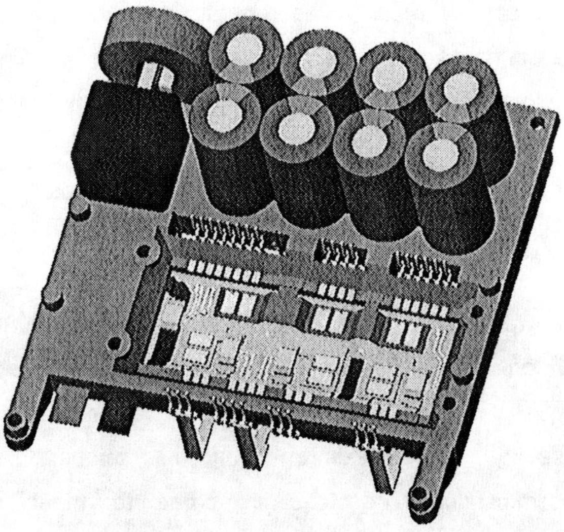

Figure 1: EPAS Module

There are numerous advantages of a pure electrical steering system versus a hydraulic system. The EPAS module is compact and modular (figure 1), and it is easily integrated and assembled into the system. These benefits are important for any kind of car, especially for smaller vehicles, where space is at a premium. Since it eliminates the need for a pump, pulley, belt, pipes network, and the fluid reservoir, it offers space and weight savings (3-5 kg). Additionally, these systems are usually sealed for life and are maintenance free. Unlike the hydraulic equivalent, in the EPAS system there are no fluids to refill, or drive belts to tighten. Furthermore, the EPAS system has a lower fuel consumption (approximately 4-5% [2]) that does not affect vehicle performance characteristics such as acceleration.

Functional advantages include a wide range of steering characteristics that can be tuned very rapidly into the vehicle. Sports and light steering settings can be programmed through software as opposed to the hydraulic version, which would require hardware alterations [3].

Unlike the hydraulic system, EPAS only draws current when it is required. Electro-Hydraulic power steering employs a pump driven by a motor, which requires continuous monitoring. A certain output pressure has to be applied at all times regardless of the need by the steering system, even when the engine is idling (i.e., compare urban driving conditions vs. freeway conditions).

These benefits, coupled with high fuel efficiency render the EPAS system as a more environmentally friendly system.

The paper focuses on the thermal analysis of EPAS power module. The EPAS module is mounted under the hood, and it must meet all of the automotive stringent quality requirements; especially, the power module has to satisfy the maximum junction temperatures of semiconductor devices in order to meet reliable continuous operation. The thermal analysis investigates the thermal response and resistance of the elements comprising the module. In addition, it allows for geometrical optimization and thermal characterization while reducing the number of iterations and cycle time.

POWER MODULE TOPOLOGY

The EPAS module is constructed with a series of layers. Each layer has either one or a combination of mechanical, electrical, or thermal functions. The primary layers are the substrate, lead frame, silicon devices, interconnects, encapsulation material and a PCB.

The power module consists of 12 dies (IRF1604 with on-state resistance of <4mΩ and break down voltage of 40V, and it is rated to junction temperature of 175^{0}C [4]). The dies are configured in a 3-phase bridge with three pairs in the electrical "low side" (closer to the motor and ground – low potential), and three pairs in the "high side" (closer to the battery – high potential). The dies are soldered to a copper lead frame on their drain side, while the source is connected to aluminum wire bonds to another part of the lead frame (drain or ground). The gate of the die is connected with an aluminum wire bond to the pins interconnects.

The lead frame that serves as a mechanical support, heat spreader, and a high-current carrier is soldered on an Insulated Metal Substrate (IMS). An IMS is a thermally efficient substrate that is made of three layers: copper (0.01 mm), dielectric (0.1 mm), and aluminum (2 mm). The effective thermal conductivity of an IMS structure is about 50 $^{W}/_{mK}$ (13% compared to copper).

A plastic shell holds the IMS, dies, and lead frame. The control circuitry is on the PCB that is mounted on the shell.

The power module is mounted on a heat sink (i.e., the chassis). In order to reduce interfacial thermal resistance, conductive thermal grease is applied between the IMS aluminum substrate and the heat sink.

THERMAL ANALYSIS DESIGN

The thermal analysis addresses three primary parameters: the high heat flux at the die footprint (60 $^{W}/_{cm^2}$), reduction of thermal stresses due to materials incompatibility, and reduction of thermal resistance. Meeting acceptable design parameters will ensure safe operating of the dies at all times. All of these parameters are addressed by thermo-mechanical analysis and testing.

Since the focus is on a steady state heat conduction mode, the analysis must identify the critical path (shortest with least amount of thermal resistance) for thermal energy to dissipate away from the dies. Identifying and optimizing this path will also reduce the time constant. Reaching steady state conditions fast is important in eliminating excessive temperature gradients that could damage the devices during transient conditions. The time constant is highly dependent on the overall system thermal mass (EPAS module, heat sink, and environment) [5].

The part was designed to operate under the worst-case scenario. The maximum heat sink (cold plate side) temperature is 105°C; however, even at these temperatures, the dies junction temperature does not exceed 175°C (lower than the maximum rating for the device).

In order to analyze the thermal behavior and validate FEA simulation, laboratory measurements were necessary. This included thermal imaging using a FLIR system, threshold voltage measurements, and thermocouple studies.

The testing required a complete electrical and mechanical test stand. The test set up included a temperature-controlled test rig with several thermocouples mounted below the dies of the power module on the top surface of a heat sink using a high thermally conductive epoxy. All thermocouples signals have been processed, and the junction temperature of each device was determined by the threshold voltage characteristics. The threshold voltage is the gate-source voltage at which an inversion channel forms. It simply represents the transition from an ON to OFF state of the device. The threshold voltage is highly dependant on the junction temperature, and therefore, it is highly reliable in determining the junction temperature. Since the threshold voltage is highly sensitive to temperature fluctuations, the measurements were taken under steady-state conditions to reduce any noise, and a dedicated data acquisition system was developed to meet this requirement. For further verifications, the FLIR system measurements were compared with the threshold voltage. Although the FLIR system measurements represent the surface temperature of the devices while the threshold voltage represent junction temperature, the difference is consolidated in the steady-state mode.

Simultaneously, a 3-D finite element model was developed using ANSYS software [6]. By combining the numerical analysis with laboratory experimentation, the design cycle has been reduced, time to market has been improved, critical areas could be addressed faster prior to production, and the number of prototypes iterations has been reduced, which resulted in further cost and time reductions during the design cycle.

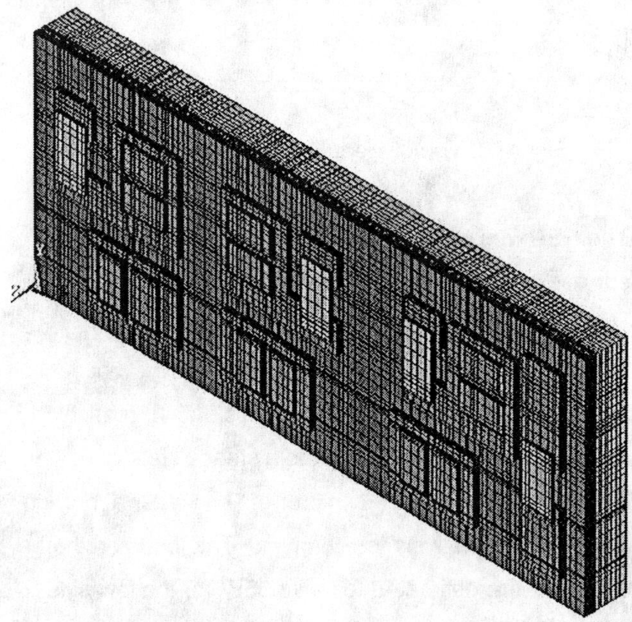

Figure 2: Meshed Power Module

During the validation stage of the finite element model, the results were compared with experimental data. Once a good correlation between numerical and experimental results was achieved, the finite element model was ready for numerical experiments and optimizations that include geometric modifications, materials selection (based on property adjustments), and variations of boundary conditions to determine module envelope of operation.

In the finite element model, the mold compound (shell), silicone gel, wire bonds, and terminal parts of the lead frames, were neglected due to their minor effect in the heat dissipation process; thus, the results obtained rendered the analysis conservative.

Once the model was built, boundary conditions and loads were applied. The finite element analysis requires data such as the thermal conductivity, density, and heat capacity, and the temperature dependence of each material used.

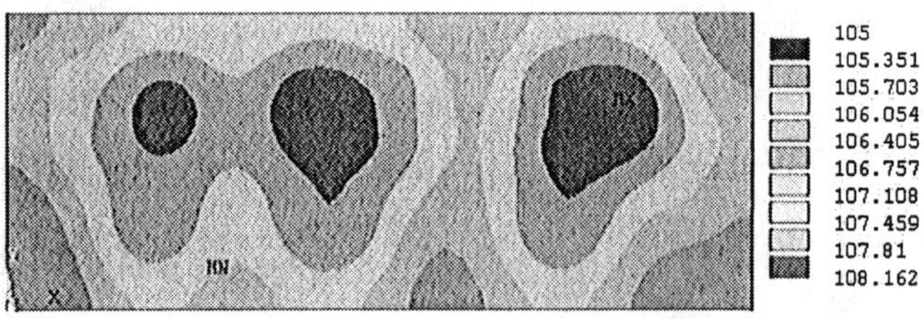

Figure 3: Top of Heat Sink Temperature Profile

The FEA was pursued conservatively and hence all surfaces were considered to be adiabatic. The steady state analysis proved most valuable in achieving a good balance between solution time and accuracy. This involved applying average power dissipation to each device until the system reached a steady state condition, instead of a real power loss profile. The latter approach rendered the analysis too complex. The loads on the devices were estimated to be 17.35W for the low-side FETs and 14.85W for the high-side FETs, with a total of 195W. Although there are differences in loads between the high- and low-side devices, the efficient thermal design through the use of the lead frame and IMS substrate (thermal conductivity and mass) reduces the junction temperature oscillation and dampens down any gradients that might cause excessive thermal stresses. This means the mechanical structure spreads heat rapidly and dampens down the junction temperature oscillations (figure 3).

Final results, shown in figure 4, demonstrate a device with a safe operating junction temperature.

The finite element model was a useful tool that was utilized in several numerical experiments that have helped to improve the power module thermal performances and efficiency, and consequently, its reliability. Ultimately, the focus of the thermal analysis was on the thermal resistance, which is used widely in the power electronic industry to characterize components thermal performance. The thermal resistance is simply defined, for the one-dimensional case, as the temperature difference between junction and case (in this case the top surface of the heat sink) over power dissipation.

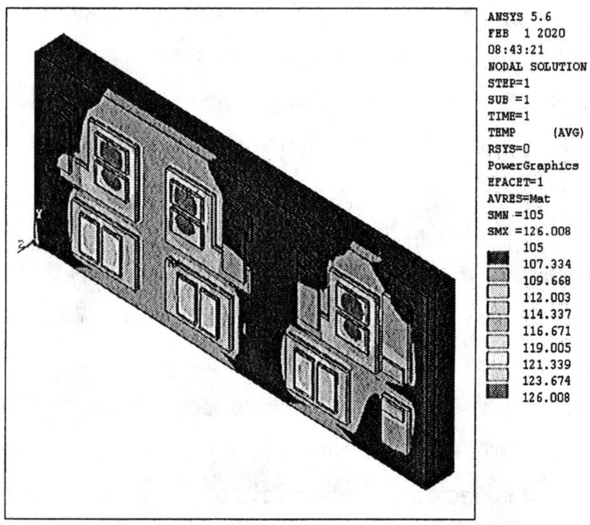

Figure 4: Steady state condition results

In order to further optimize the thermal resistance, the numerical model was used to identify the thermal resistance contributions of each of the power module layers. The graph on Figure 5 shows the temperature gradient through the power module, from the top of the dies down to the heat sink. The temperature variation through each single layer is shown, so that it is easy to identify critical areas of improvement that experience the most thermal resistance (inhibits thermal energy flow).

In this case, it is clear that the two dielectric and thermal grease layers are the main contributors to the thermal resistance. Maximizing the layers thermal conductivity by employing dielectrics with improved characteristics and optimizing their thickness has improved the thermal resistance.

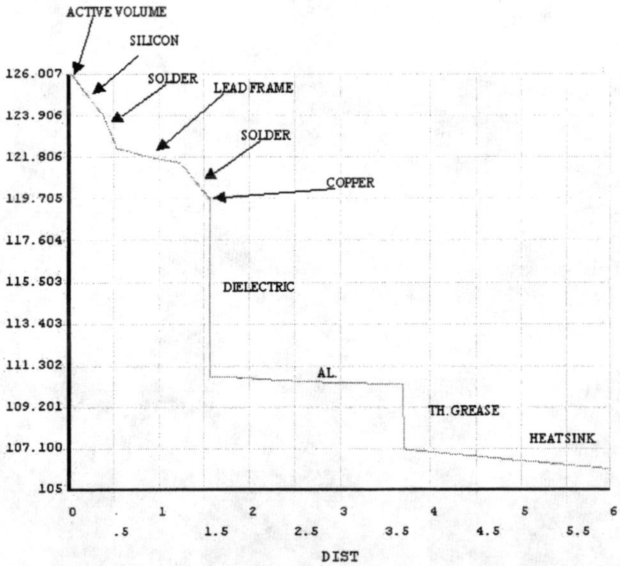

Figure 5: Layer Thermal Profile

CONCLUSION

The EPAS package has been designed to perform reliably in the harsh under the hood conditions. The power module has a power density of $20W/_{cm^{3}}$ and thermal resistance of $1.2\ ^{W}/_{C}$., and the dies reach a peak junction temperature of less than $130^{0}C$. This is primarily due to the IMS substrate and lead frames that function as a heat spreader. In fact, the heat flux is reduced from a maximum of $60\ ^{W}/_{cm^{2}}$ down to $7.5\ ^{W}/_{cm^{2}}$ (a factor of 8). The encapsulation materials keep lead frames and a PCB in place and preserve silicon devices from being contaminated, while absorbing some of the thermal energy. The control circuitry is located on the PCB.

Worst-case scenarios have been analyzed using finite element method and laboratory verification. Once the finite element model was validated, several numerical simulations of EPAS power module behavior were performed to optimize the thermo-mechanical design, verifying that silicon devices junction temperatures were lower than specified limits ($175^{0}C$).

FEA makes possible the reduction of time to market and product development cost reductions, while the geometrical topology adopted for this module has been optimized for a high thermal efficiency of 86%. This thermal efficiency represents the ratio of the difference between the power input and losses to the power input.

Due to advances in semiconductor technology, the trend with silicon devices is to become smaller and more efficient (i.e. lower power dissipation). Thus, the future generations of this module will be even more compact and lighter, with an improved thermal efficiency characteristic.

ACKNOWLEDGMENTS

I would like to thank all people who have participated in the EPAS power module development. In particular, I would like to thank Steve Clemente, Bill Grant, and Sergio Fissore for advising and counseling throughout the project.

REFERENCES

[1] C. Webber, ed. “Automotive Semiconductor Forecast 1998-2007,” Strategy Analytics. September 2000.

[2] Lidow Alex, "The Power Conversion Process as a Prosperity Machine", International Rectifier, http://www.irf.com, 1999.

[3] S. Fissore and L. Cormie, "Development of Power Assisted Steering Package." SAE, *Transitioning to 42V Electrical System Conf.* pp. 103-108 (2000).

[4] IR1F1604 Datasheet, International Rectifier, http://www.irf.com.

[5] Mills, Anthony F., "Heat & Mass Transfer", IRWIN, Chicago, 1995.

[6] *ANSYS Technical overview*, USA, SAS IP, Inc., 1997.

DEFINITIONS, ACRONYMS, ABBREVIATIONS

FET Field Effect Transistor

EPAS Electronic Power Assisted Steering

CAGR Cumulative Annual Growth Rate

FEA Finite Element Analysis

FEM Finite Element Method

FLIR Forward Looking Infrared

PCB Printed Circuit Board

Proceedings of
2001 ASME International Mechanical Engineering Congress and Exposition
November 11–16, 2001, New York, NY
EPP-Vol. 1

IMECE2001/EPP-24713

NONINVASIVE MEASUREMENTS AND COMPUTATIONAL MODELING OF SMT ASSEMBLIES

Ryszard J. Pryputniewicz, David Rosato*, and Cosme Furlong
NEST – NanoEngineering, Science, and Technology
CHSLT – Center for Holographic Studies and Laser micro-mechaTronics
Mechanical Engineering Department
Worcester Polytechnic Institute
Worcester, MA 01609

* Harvard Thermal, Inc.
249 Ayer Road, Suite 102
Harvard,

ABSTRACT

Differential thermal expansions cause strains in the interconnection structures, during functional operation of active devices. To evaluate these strains, temperature profiles of the interconnected components must be known. In this paper, a methodology for developing thermal models of surface mount technology (SMT) assemblies is presented using thermal analysis system (TAS) and its application is demonstrated by simulating thermal fields of a representative package. Thermomechanical deformations of the package are measured quantitatively using state-of-the-art laser-based optoelectronic holography (OEH) methodology. Results of TAS simulation show good correlation with noninvasive full-field-of-view OEH measurements of thermomechanical field effects of the SMT assembly under actual operating conditions.

Keywords: SMT, thermal analysis system, thermal models of SMT packages, optoelectronic holography, noninvasive measurements of thermal deformations.

1. INTRODUCTION

Surface mount technology (SMT) has become the mainstream packaging technology of electronic equipment manufacturers. The significant driving force for the use of SMT is the reduced package size and cost, as well as improved board utilization as the package-to-board interconnection spacing is minimized, when compared to pinned-through-hole packages. However, placing more functions in a smaller package has resulted in higher heat densities, which require that thermal management be given a high priority in design so as to maintain system performance and reliability.

As active devices dissipate heat to their surroundings, or as the ambient system temperature changes, differential thermal expansions cause stresses in the interconnection structure. These stresses produce elastic and plastic strains, as well as time-dependent deformations of the joints between the relatively rigid package body and circuit board. This becomes important as the devices are thermally cycled during operation, given the differences in thermal expansion for the materials of package and board, and given the differences in temperature between the package and the board.

Knowledge of thermal-cycle-induced characteristics of SMT joints is important. This paper describes preliminary results of the study using noninvasive measurements and computational modeling methodologies for evaluating thermomechanical effects in the SMT assemblies.

2. SMT ASSEMBLY

The SMT assembly used in this preliminary study is shown in Fig. 1. More specifically, SMT packages attached to a board via gull wing leads, Fig. 2, were evaluated for thermal fields and deformations during their functional operation.

3. METHODOLOGIES USED

Thermomechanical effects in the SMT assemblies were studied, in this paper, using a computational methodology based on the thermal analysis system (TAS) and an experimental methodology utilizing a laser-based optoelectronic holography (OEH).

Fig. 1. SMT assembly used in this study.

Fig. 2. Detail of the SMT configuration.

3.1. Thermal analysis system

To evaluate the magnitude of thermally induced deformations in a system, the temperature profile of the interconnected components comprising the system must be known throughout an operational cycle or system temperature excursion. In this study, the temperature profiles were determined using thermal analysis system (TAS) model based on a resistor network. TAS contains a finite difference solver, which performs heat balance at each node of the model. This entails calculating the node temperature based on the resistances and the temperatures of all nodes attached to the node in question.

3.1.1. Resistor network

The resistor network is obtained by converting all elements into resistances [1,2]. Values of the resistances are functions of the geometry of the elements and the thermal conductivity. For example, for brick elements whose internal elements are all 90° and whose thermal conductivity is isotropic, 12 resistors, denoted by R_k, are calculated according to the equation

$$R_k = \frac{\Delta}{kA_k} \quad , \tag{1}$$

where Δ is the distance between the nodes, k is the thermal conductivity of the element, and area A_k is the cross-sectional area for a portion of the element associated with the interconnected nodes. This is calculated as the product of half the width of the element normal to the heat flow in both directions for the specific resistor. If the internal angles of a brick element are not all 90°, or the thermal conductivity is anisotropic, then the brick element is represented by 28 resistors.

A similar method is used for wedge shaped elements, which is represented by 15 resistances.

Resistance R_c of a convection element, which defines convective surfaces, is calculated using the equation

$$R_c = \frac{1}{hA_c} \quad , \tag{2}$$

where h is the convection heat transfer coefficient and A_c is the convective area associated with each note on a surface.

For heat to be transferred between elements of the model, they must share nodes. Because of this, in the process of calculating resistances, many resistors are generated between the same nodes. The model could be solved at this point, but the solution would be inefficient. To speed up the solution, resistors in the model that go between the same nodes are combined. If the resistances are fixed, i.e., if they are neither time nor temperature dependent, or are time and temperature dependent and reference the same array, they are combined into a single resistor using the following equation:

$$R = \left(\frac{1}{R_i} + \frac{1}{R_{i+1}} + \cdots + \frac{1}{R_{i+n}} \right)^{-1} \quad . \tag{3}$$

Application of Eq. 3 takes only a few seconds on a very large model, but can reduce the size of the model by a factor of 3 or 4.

Resistances of surface radiation elements, R_r, are calculated from each node of the element face to a single reference node according to the equation

$$R_r = \frac{1}{K_r \left(T_1^2 + T_2^2\right)\left(T_1 + T_2\right)} \quad , \tag{4}$$

where T_1 and T_2 are the absolute temperatures of the two nodes to which the radiation resistor is connected, while

$$K_r = A_r \sigma F \varepsilon \quad . \tag{5}$$

In Eq. 5, A_r defines the radiation area associated with each node of the face geometry of the specific element, σ is the Stefan-Boltzmann constant, F is the view factor, and ε is the emissivity.

The radiation resistances are calculated every iteration during the solution since their values depend upon the nodal temperatures. Having radiation in the network makes the model nonlinear.

3.1.2. Details of the TAS solution

All resistor data are streamlined to result in a fast running solution, which is facilitated by two finite difference routines used for: (1) steady state solution, and (2) transient solution.

The finite difference routine used for the steady state solutions is

$$T_i = T_{i_{prev}}(1-d) + d\frac{\sum\left(\frac{T_{i+1}}{R_{i+1}}\right) + Q_i}{\sum\left(\frac{1}{R_{i+1}}\right)} \quad , \tag{6}$$

where T_i is the new calculated temperature of the node for which the heat balance is being performed, $T_{i_{prev}}$ is the temperature of the node from the previous iteration, T_{i+1} is the temperature of the node connected to the node T_i by the resistor R_{i+1}, and Q_i is the total heat load on the node for which the heat balance is being performed – this may be a fixed value or a temperature dependent value.

In Eq. 6, the summations are for all resistors attached to the node for which the heat balance is being performed, while d is the damping coefficient with the default value of one, i.e., no damping. If the routine senses that the solution is diverging rather then converging, the damping coefficient is reduced by 0.05. This continues, during each iteration, until the solution begins to converge. If severe diverging is sensed then the damping coefficient is immediately given a value of 0.25. It should be realized that damping is only needed in some nonlinear solutions, e.g., ones involving radiation, or variable material properties.

The finite difference routine used for transient solutions is

$$T_i = \frac{\left[\sum\left(\frac{T_{i+1} - T_i}{R_{i+1}}\right) + Q_i\right]\Delta t}{\rho c_p V} + T_{i_{prev}} \quad , \tag{7}$$

where Δt is time increment, the product $\rho c_p V$ represents the thermal capacitance of the node, and other variables are as defined previously.

During TAS model execution, the array data are interpolated for each computation time step. Representative results are shown in Section 4.

3.2. Optoelectronic holography

Optoelectronic holography (OEH) is the state-of-the-art laser-based methodology for rapid (i.e., near real-time), noninvasive, full filed of view, quantitative studies of deformations of structures with high spatial resolution and high measurement accuracy [3-6]. Figure 3 shows a typical OEH system used for quantitative studies of thermal deformations of SMT. This system, using fiber optics to guide the laser beam, is configured in a single-illumination and single-observation geometry.

In the OEH system shown in Fig. 3, laser light is coupled into a single fiber optic by means of a microscope objective (MO). Then, the single fiber is divided into two parts by means of a directional coupler (DC) which produces two laser beams. One of the laser beams is used to illuminate the object, while the other beam provides reference against which the signals from the object are recorded. Both, the object and reference beams are combined by the interferometer (IT) and recorded by the system camera (CCD).

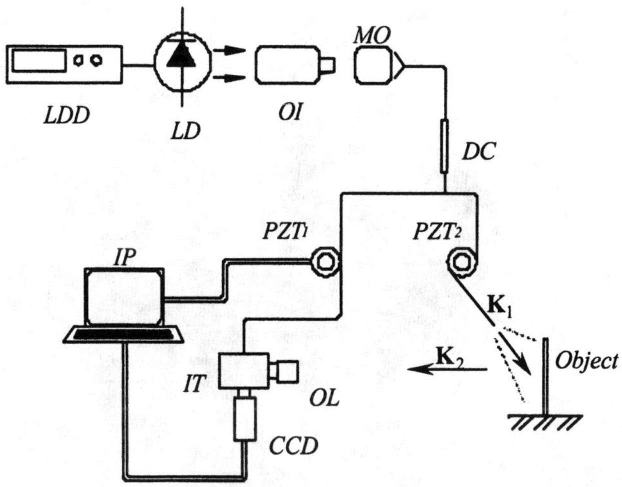

Fig. 3. Single-illumination and single-observation geometry of a fiber optic based OEH system for applications in electronic packaging: *LDD* is the laser diode driver, *LD* is the laser diode, *OI* is the optical isolator, *MO* is the microscope objective, *DC* is the fiber optic directional coupler, PZT_1 and PZT_2 are the piezoelectric fiber optic modulators, *IP* is the image processing computer, *IT* is the interferometer, *OL* is the objective lens, *CCD* is the camera, while $\mathbf{K}_1$ and $\mathbf{K}_2$ are the directions of illumination and observation, respectively.

The images recorded by the CCD are processed to determine the fringe-locus function, $\Omega(x,y)$, constant values of which define fringe loci on the surface of object under investigation [7-8]. The values of $\Omega(x,y)$ relate to the system geometry and the unknown deformations via the relationship

$$\Omega(x,y) = [\mathbf{K}_2(x,y) - \mathbf{K}_1(x,y)] \bullet \mathbf{L}(x,y) = \mathbf{K}(x,y) \bullet \mathbf{L}(x,y) \quad , \tag{8}$$

where $\mathbf{K}$ is the sensitivity vector defined in terms of vectors $\mathbf{K}_1$ and $\mathbf{K}_2$ identifying directions of illumination and observation in the OEH system, Fig. 3, $\mathbf{L}$ is the deformation vector, and the argument (x,y) defines spatial coordinates x and y.

Quantitative determination of structural deformations due to thermal fields induced by electric loads can be obtained, by solving a system of equations similar to Eq. 8, to yield [7-10]

$$\mathbf{L}(x,y) = [\mathbf{K}(x,y), \Omega(x,y)] \quad . \tag{9}$$

Representative results obtained by solution of Eq. 9 are discussed in Section 4.

4. RESULTS AND DISCUSSION

An example of a component considered in this study is shown in Fig. 4. Its TAS model, shown in Fig. 5, displays the level of detailed thermal-resistance modeling of the device, package, joint, and board assembly that is necessary to predict temperature distributions. Once the thermal model for a particular package is generated, it can be applied to the specific component/board environment to be evaluated. The result is a set of temperature profiles for the package and environments, as indicated in Fig. 6.

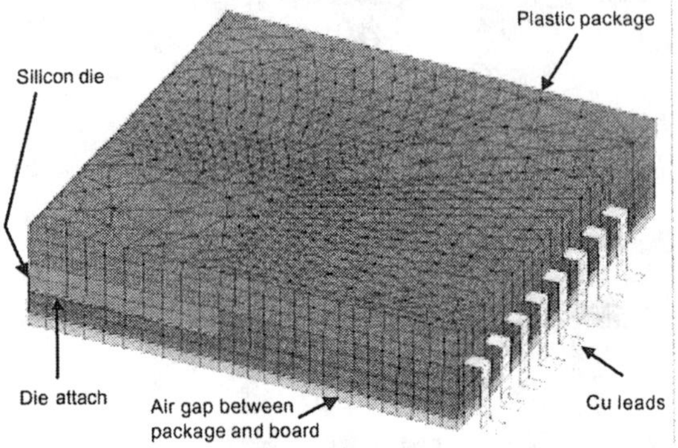

Fig. 4. Description of the SMT component considered in this study.

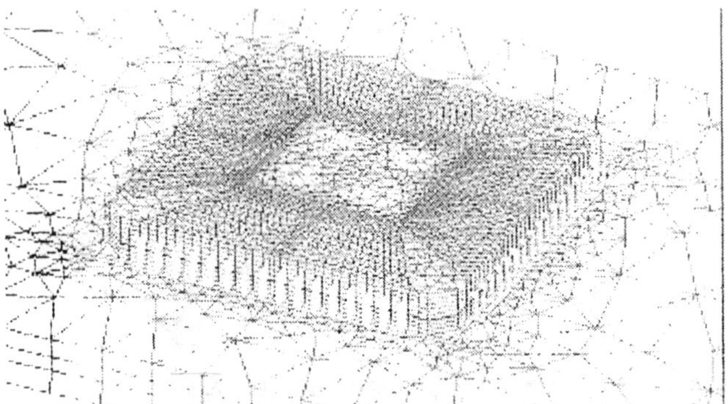
Fig. 5. TAS model of the component shown in Fig. 4.

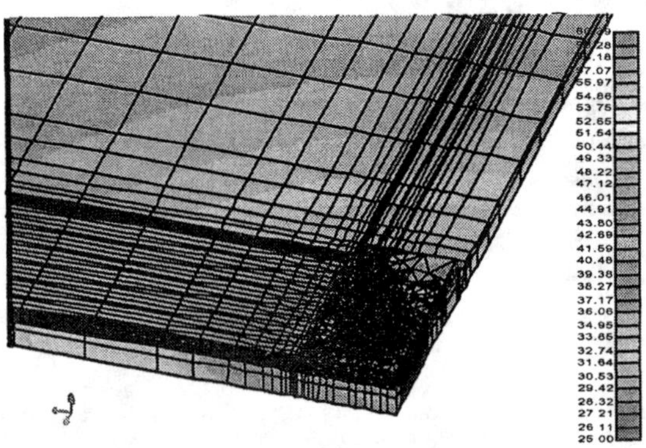
Fig. 6. TAS computed temperature distribution for the model shown in Fig. 5. Because of the symmetry, quarter-model is shown.

Once the time-dependent temperature profile is known, the thermal deformations may be predicted for the rigid elements of the structure.

OEH fringe pattern, encoding thermal deformations due to power on cycle, of the SMT component is shown in Fig. 7. Figure 8 indicates the maximum deformation of 1.87 μm, based on the fringe pattern of Fig. 7, while Fig. 9 displays wireframe representation of the 3D deformations corresponding to Figs 7 and 8.

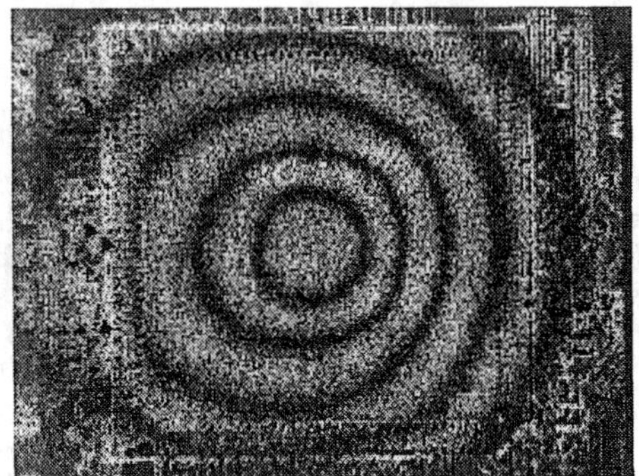
Fig. 7. OEH fringe pattern of thermal deformations due to power on of the SMT component.

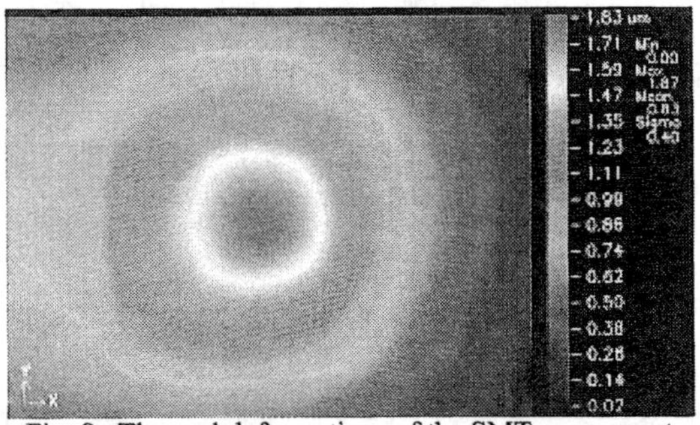

Fig. 8. Thermal deformations of the SMT component, based on the OEH fringe pattern of Fig. 7.

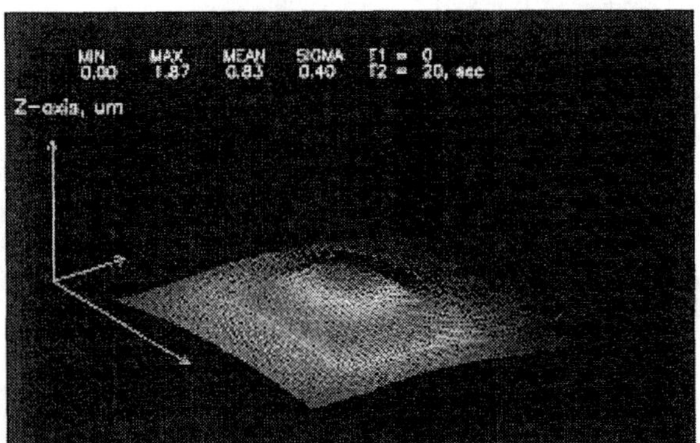

Fig. 9. Wireframe representation of thermal deformations of the SMT component based on the OEH fringe pattern of Fig. 7.

Comparing Figs 6 and 8, similarity in temperature profiles and deformation patterns, respectively, is noted.

The OEH fringe pattern of the SMT configuration shown in Fig. 2 is displayed in Fig. 10. This pattern is quantitatively analyzed to obtain results of Figs 11 and 12, which indicate that the maximum deformation is 3.19 μm.

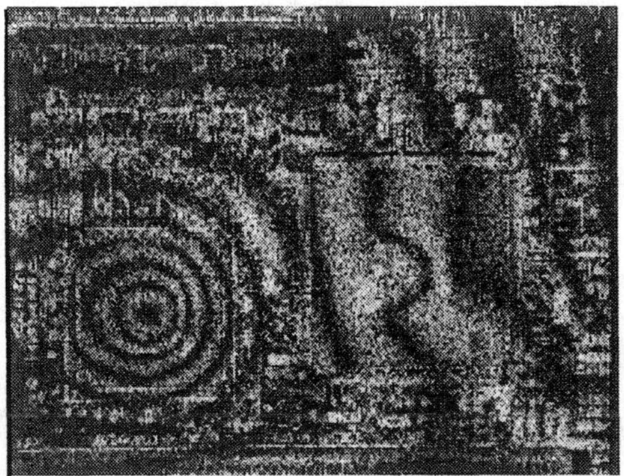

Fig. 10. OEH fringe pattern of the SMT configuration shown in Fig. 2

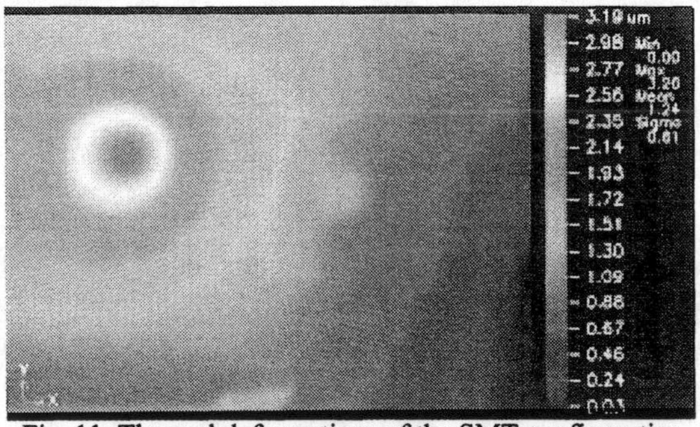

Fig. 11. Thermal deformations of the SMT configuration based on the OEH fringe pattern of Fig. 10.

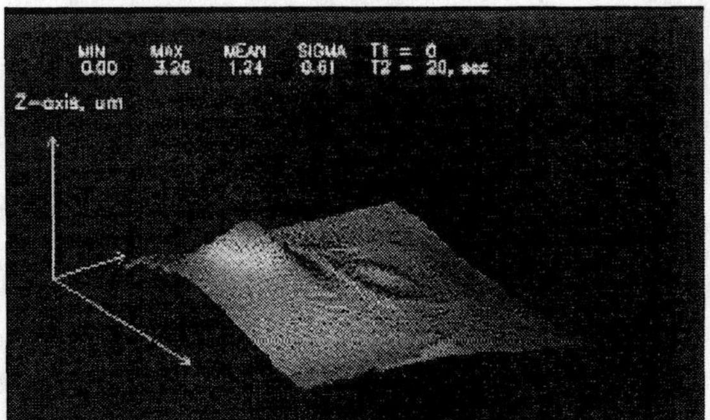

Fig. 12. Wireframe representation of thermal deformations of SMT component based on the OEH fringe pattern of Fig. 10.

5. CONCLUSIONS

Noninvasive methodology for quantitative modeling and measurements of SMT assemblies was presented. This methodology consists of thermal analysis system (TAS) and laser-based optoelectronic holography (OEH) method. TAS contains a finite difference solver, which performs heat balance at each node of the model consisting of a resistance network. OEH performs remote, full-field-of-view measurements of deformations of the SMT assemblies. These measurements are performed in near real-time, with high spatial resolution and nanometer measurement accuracy.

In this study, modeling and measurements of SMT assemblies was performed. The assemblies used consisted of a single component and two component configurations.

Preliminary results indicate that during power-on cycle, maximum deformations of a single component are on the order of 1.87 μm. The results also indicate that the maximum deformations of the SMT assembly, consisting of a two component configuration, are on the order of 3.26 μm.

Comparison of data obtained using TAS and OEH methods indicated good correlation between the computational and experimental results.

The preliminary results shown in this paper indicate that hybridization of TAS and OEH methods provides a useful new tool for quantitative studies of SMT assemblies.

Our work on this noninvasive methodology for quantitative modeling and measurements of SMT assemblies will continue.

6. REFERENCES

1. Rosato, D., 1988, *Thermal analysis system*, User's guide, Harvard Thermal, Inc., Harvard, MA.
2. Kreith, F., 1968, *Principles of heat transfer*, International Textbook Co., Scranton, PA.
3. Pryputniewicz, R. J., 1985, "Heterodyne holography applications in studies of small components," Opt. *Eng.*, 24:849-854.
4. Pryputniewicz, R. J., 1994, "A hybrid approach to deformation analysis," *Proc. SPIE*, 2342:282-296.
5. Furlong, C., 1999, Hybrid, experimental and computational, approach for the efficient study and optimization of mechanical and electro-mechanical components, Ph.D. Dissertation, Worcester Polytechnic Institute, Worcester, MA.
6. Furlong, C., and Pryputniewicz, R. J., 2001, "Sensitivity, accuracy, and precision issues in quantitative optical metrology characterizations," *Proc. Internat. Congress on Experimental and Applied Mechanics for Emerging Technologies*, SEM, Bethel, CT, pp. 631-634.
7. Pryputniewicz, R. J., and Furlong, C., 2000, "Novel optical-computational approach for NDT applications in microelectronics," *Proc. IX Internat. Congress on Experimental Mechanics*, SEM, Bethel, CT.
8. R. J. Pryputniewicz, 1992, *Holographic numerical analysis*, Worcester Polytechnic Institute, Worcester, MA.
9. R. J. Pryputniewicz, 1981, "High precision hologrammetry," *Internat. Arch. Photogramm.*, 24:377-386.
10. C. Furlong and R. J. Pryputniewicz, 2000, "Absolute shape measurements using ultra-high resolution optoelectronic holography methods," *Opt. Eng.*, 39:216-223.

Proceedings of
2001 ASME International Mechanical Engineering Congress and Exposition
November 11–16, 2001, New York, NY
EPP-Vol. 1

IMECE2001/EPP-24714

SILICON CARBIDE WAFER POLISHING WITH GAS CLUSTER ION BEAMS

V. DiFilippo[1], J.A. Bennett[2], D.B Fenner[3], J.K. Hirvonen[4], L.C. Feldman[2], A. Saigal[1]

[1]Tufts University, Mechanical Engineering Department, Medford, MA 02155
[2]Vanderbilt University, Department of Physics and Astronomy, Nashville, TN 37235
[3]Epion Corporation, Billerica, MA 01821
[4]Army Research Laboratory, Aberdeen Proving Ground, MD 21005

ABSTRACT

A novel technology utilizing energetic ionized gas cluster ion beams (GCIB) has been successfully used to reduce the surface roughness of SiC for electronic applications. In GCIB, a high-pressure gas, such as argon, is supersonically expanded through a nozzle into vacuum. This adiabatic expansion results in the condensation of clusters consisting of tens to thousands of gas atoms weakly held together by Van der Waals forces. These clusters are then ionized and accelerated towards the target substrate. Upon impact, they create a strong lateral sputtering effect resulting in a net smoothing of the surface. The surfaces of commercial, single-crystal, research grade SiC wafers were exposed to a series of Ar and O_2 gas cluster ion treatments of varying doses and energies. After processing, atomic force microscope measurements indicated a reduction of surface roughness and removal of pre-existing CMP polishing scratches. Enhanced GCIB smoothing was achieved by using sequential doses of gas clusters with different energies, and in some cases different gases. Higher cluster energies removed scratches but added a high spatial-frequency roughness component, which could be in turn reduced by lower energy clusters. It was found that oxygen clusters produced a greater degree of smoothing than argon clusters. RBS ion channeling measurements were also conducted, and showed that processing with lower energy clusters as the final step of a series of doses minimized the creation of defects near the surface.

INTRODUCTION

SiC is being considered for use in electrical switches and other electronic devices operating at high currents, power densities and temperatures. A major factor limiting the commercialization of these devices is the high cost of SiC substrates. A potential solution is the use of ion beam cleaving to transfer thin SiC layers to less expensive substrates. In one version of ion beam cleaving co-developed at Vanderbilt University and ORNL, H^+ and Si^+ ions are implanted into the substrate to induce blister formation and layer separation [1]. Unfortunately, this process results in a rough surface topography, which limits device performance. Traditional polishing techniques such as chemical etching are not able to produce a satisfactory surface. Additionally, polishing capability is limited by the <1 μm thickness of the transferred layer. In response to this problem a novel technology utilizing energetic ionized beams of gas clusters (GCIB) has been used to improve the surface roughness of SiC [2].

GCIB is a relatively new technique whereby energetic ionized clusters of condensed gas atoms, typically Ar, are directed towards the target. These clusters of tens to thousands of gas atoms or molecules are weakly held together by van der Waals forces. They are formed by homogeneous condensation occurring within the flow of a high-pressure gas expanding into a vacuum. Neutral clusters thus formed are then ionized and accelerated through a high voltage potential, producing an energetic beam of cluster ions. Cluster ion beams have markedly different effects on the target material than do conventional monomer ion beams. Although the cluster ions are highly energetic (10 – 30 keV acceleration), this energy is shared between many atoms in the cluster giving each individual atom an average energy of 5-50 eV. This results in very little penetration and therefore a higher concentration of energy within a smaller volume. This produces much higher local impact temperatures and pressures than monomer ions as well as higher sputtering yields. Consequently, GCIB offers effective surface smoothing effects. GCIB has been used to produce exceedingly smooth surfaces with an average roughness of as low as two angstroms [3-12].

The unusual processing effects that can be produced by cluster ions are related to the combination of high total energy, mass and momentum but low energy per atom in the clusters. In addition, there are other factors that distinguish the impacts of gas cluster ions upon solid surfaces from corresponding impacts of monomer ions. These include the influence of virtually simultaneous impact of the target surface by a large number of spatially coincident atoms, and the manner in which atoms of the clusters interact with each other as the clusters disintegrates [8].

A significant smoothing and planarization effect is produced during gas cluster ion impacts on nonplanar surfaces. As atoms within a cluster are able to interact with each other as the cluster disintegrates upon impact, some of the total energy carried by the cluster is converted to individual atom energy directed along the plane of the target surface. The smoothing produced by gas cluster ion impacts is somewhat analogous to a balloon full of marbles thrown onto a plane surface under conditions whereby the balloon breaks and the marbles become scattered in all directions over the target due to interactions among themselves during the impact. This lateral sputtering effect interacts with surface features, preferentially etching peaks, to cause smoothing and planarization. Gas cluster ions have been experimentally demonstrated to produce atomic scale surface smoothing on virtually all solid inorganic materials. Examples include metals [Ta, Cu, Au, Ag], semiconductors and metal oxide superconductors such as $YBa_2Cu_3O_{7-\delta}$ [6-12].

EXPERIMENTAL PROCEDURE

Figure 1 illustrates the gas cluster ion beam processing system at the Epion Corporation used in this investigation. Gas is expanded in the nozzle and clusters are formed. The core of the beam enters the ionization section through a narrow skimmer where the clusters are ionized and then accelerated towards the target by an acceleration voltage. A permanent magnet deflects the lighter monomer species out of the beam, allowing only clusters to reach the target. Electrostatic scanning plates raster the beam to ensure a uniform dosage. The SiC wafer was covered with a tantalum aperture to further ensure uniform beam strength over the process area and to allow precise measurements of beam current and ion dosage. Typical processing chamber base pressure was $1x10^{-7}$ torr.

Two, electronic grade, SiC wafers were used. One wafer (A) was processed with only Ar clusters while the second wafer (B) was polished with both Ar and O_2 clusters. A series of ion doses at different cluster energy levels was applied to separate process areas on each wafer. On the second wafer in particular, a number of process areas were exposed to two consecutive ion doses in which the cluster energy and gas were varied. Cluster dose and cluster energy were the two GCIB process parameters that were varied. Cluster dose was controlled by controlling the exposure time to the cluster beam, while adjusting the acceleration voltage controlled the cluster energy. Wafers A and B were exposed to 15 distinct conditions, the doses and acceleration voltages for the most relevant of these are listed in Table 1 [13].

Surfaces were measured with a Digital Instruments atomic force microscope (AFM) equipped with analysis software, that was used to calculate surface roughness values such as arithmetic mean roughness (R_a) and maximum peak to valley height (R_{max}). Rutherford backscattering (RBS) channeling measurements were also performed on the wafers to measure the extent of near surface crystal lattice damage. Ion channeling measurements were performed with He^+ ions at 1.8 MeV after etching with hydrofluoric acid to remove any oxide layer.

Figure 2 shows a 5µm x 5µm AFM scan, which is typical of the surfaces of the SiC wafers. This surface distinctly shows the effects of mechanical polishing. The large number of surface scratches should be noted. The more pronounced scratches are on the order of 50 Å deep. As indicated in Table 1, an average of several measurements over both wafers found the virgin surface roughness to have an R_a of 8.4 Å for wafer A and 9.6 Å for wafer B.

RESULTS

As described above and shown in Figure 2, many polishing scratches are visible on the as-received from vendor surface of the SiC wafers. As shown in Table 1, condition A4, a single dose of Ar clusters at a moderate (15 KeV) cluster energy failed to improve the measured surface roughness.

Figure 3 shows the effect of GCIB processing on the SiC wafer surface with O_2 clusters. At low cluster energy (condition B4 in Table 1) the polishing scratches are less distinct. At a higher energy (condition B1) the polishing scratches are completely gone, but a higher frequency roughness is now present. Figure 3 and Table 1 show that the low energy treatment slightly lowered the roughness while the high-energy process increased it.

A plot of R_a normalized to the untreated roughness value versus ion dose is given in Figure 4. It shows the effect on surface roughness as a function of ion dose for 1) Ar single step processes and 2) Ar and O_2 dual energy processes. In the case of Ar alone, there is an initial is an increase in roughness from 8.4 Å (as received) to 11 Å and then a gradual smoothing with increasing dose. Table 1 shows a large decrease in R_{max} values for Ar processing, from 192 Å on the untreated surface down to 105 Å (condition A3).

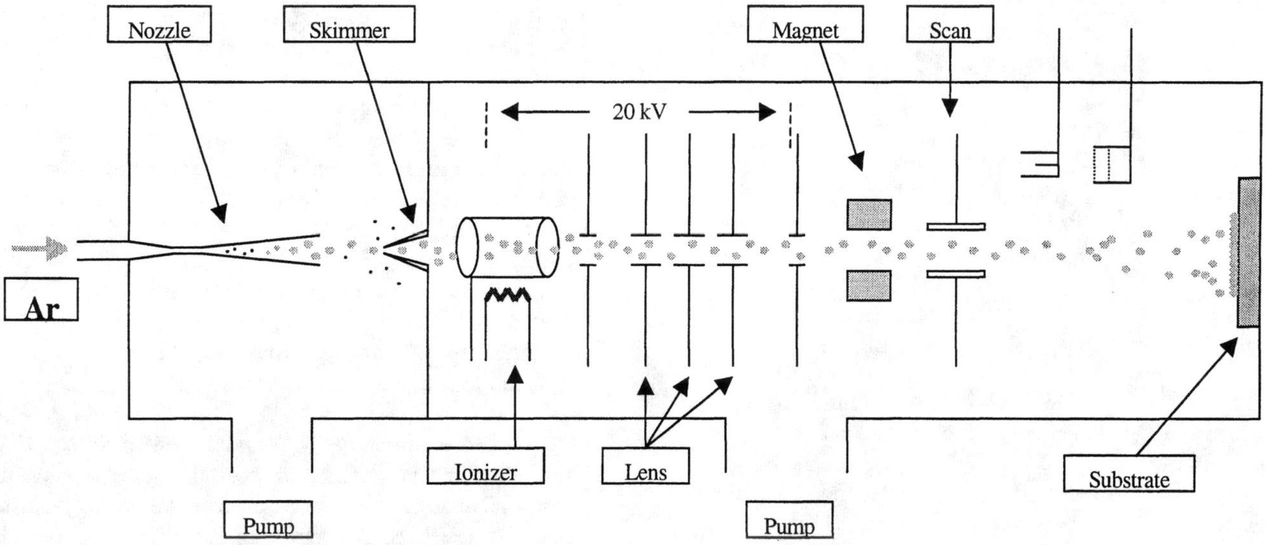

Figure 1: GCIB System

Table 1: AFM and RBS of SiC Wafers

#	GCIB Gas	Accel (kV)	Dose (ions/cm^2)	R_a (Å)	R_{max} (Å)	N_{sp} ($1x10^{16}$ atoms/cm^2)	χ_{min} (%)
Virgin A	- -	- -	- -	8.4 ±0.5	192 ±23	0.67	2.3
A1	Ar	25	$1x10^{15}$	11.0	203	3.24	4.6
A2	Ar	25	$3x10^{15}$	9.5	178	2.68	4.6
A3	Ar	25	$1x10^{16}$	8.2	105	3.05	4.5
A4	Ar	15	$1x10^{16}$	16.0	210	3.19	5.1
A5	Ar, Ar	25, 15	$1x10^{16}$, $1x10^{16}$	7.0	86	1.60	4.2
Virgin B	- -	- -	- -	9.6 ±0.9	248 ±22	0.63	2.4
B1	O_2	25	$1x10^{16}$	11.6	355	0.99	2.5
B2	Ar, O_2	25, 7	$1x10^{16}$, $1x10^{15}$	5.1	72	3.78	5.6
B3	O_2, O_2	25, 7	$1x10^{16}$, $1x10^{15}$	3.8	69	0.81	2.4
B4	O_2	7	$1x10^{15}$	9.1	157	0.64	1.9

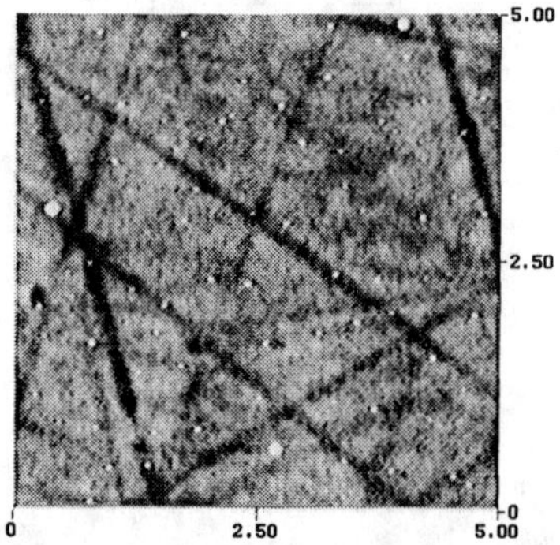

Figure 2: 5μm x 5μm AFM image of untreated SiC wafer B surface. (R_a = 9.6 Å, R_{max} = 248 Å)

The greatest reduction in R_a and R_{max} while utilizing only Ar was obtained using a dual energy smoothing process consisting of an initial dose at high energy followed by a dose at a moderate energy (condition A5). As shown in Figure 4, this dual energy process produced a greater reduction in surface roughness than an equivalent dose using only high energy clusters. The result was a final R_a of 7.1 Å and R_{max} of 86 Å.

Processing with O_2 clusters produced similar results. Low energy oxygen cluster doses alone did not significantly reduce surface roughness or remove polishing scratches (condition B4, Figure 3a). High energy oxygen doses were needed to alter the surface morphology and remove scratches (condition B1, Figure 3b). As with Ar clusters, these high-energy oxygen doses introduced a high frequency roughness component, which needed to be reduced by a subsequent lower energy oxygen dose. The result of such a dual energy oxygen process (condition B3) is shown in the AFM scan of Figure 5. As with the argon dual energy process both low frequency polishing scratches and high frequency roughness were reduced. However, the oxygen process produced an overall smoother surface than the argon process with a final R_a of 3.8 Å and R_{max} of 69 Å.

Defect density (N_{sp}) and Chi Min values (X_{min}) from RBS measurements of crystal lattice damage are given in Table 1. Areas exposed to lower energy clusters had lower levels of damage than those exposed to higher energy clusters. The areas exposed to a dual energy process (high energy dose followed by lower energy dose) had still lower damage levels, similar to the unprocessed areas. Regions processed with only oxygen clusters had significantly lower damage levels than those processed with argon.

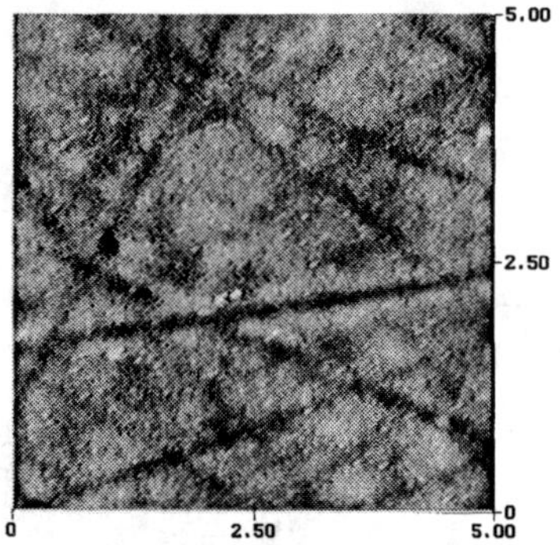

(a)

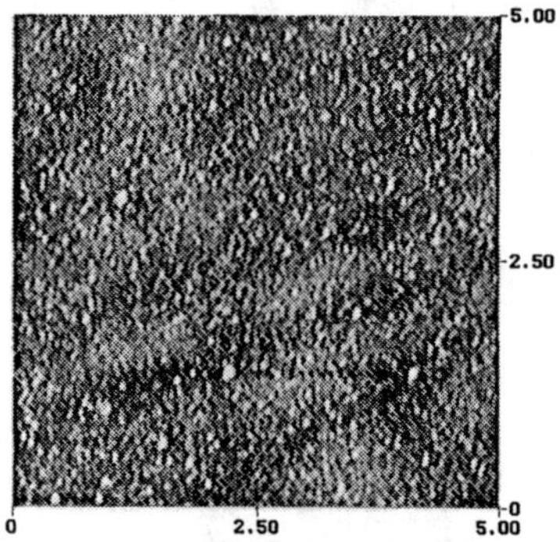

(b)

Figure 3: 5μm x 5μm AFM images of SiC surfaces exposed to different GCIB oxygen cluster energies: a) low energy (condition B4), R_a = 9.1 Å, R_{max} = 157 Å; b) high energy (condition B1), R_a = 11.6 Å, R_{max} = 355 Å

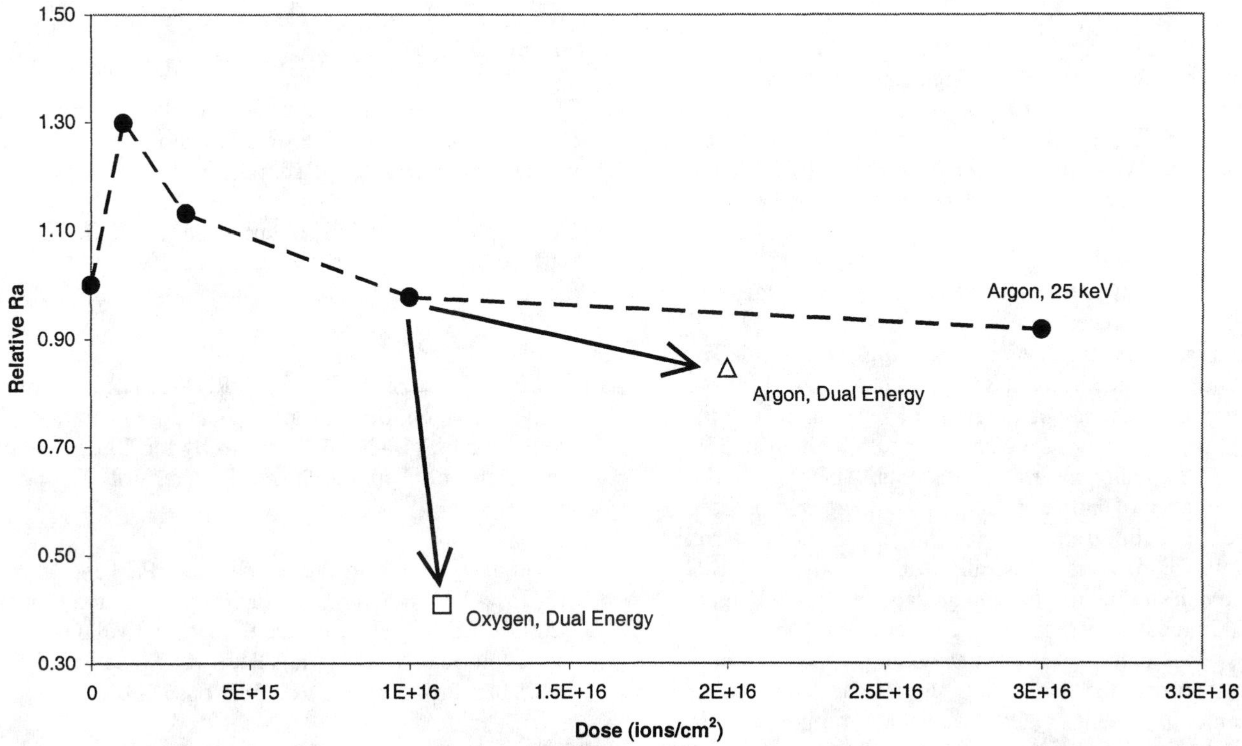

Figure 4: Plot of relative R_a (relative to as received roughness) vs. dose at high cluster energy (25 kV) with Ar clusters. Also shown is the effect of a dual energy smoothing process, consisting of a dose at high energy followed by a dose at lower energy for O_2 (condition B3) and Ar (condition B2). (The lines are only a guide to the eye)

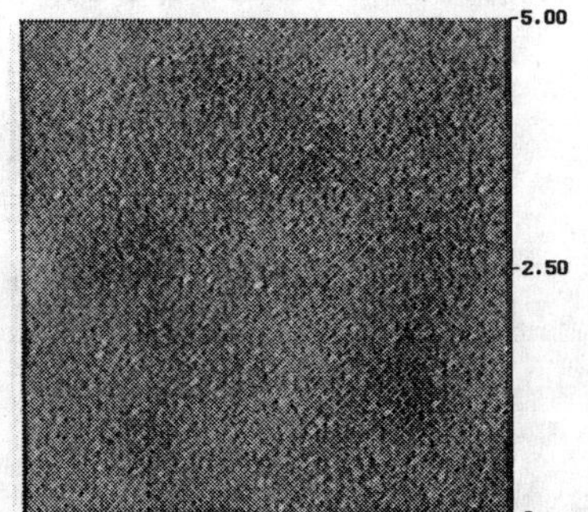

Figure 5: 5µm x 5µm AFM images of SiC surfaces exposed to GCIB polishing with O_2 clusters at two energy levels. (R_a = 3.8 Å, R_{max} = 69 Å, condition B3)

DISCUSSION AND CONCLUSIONS

The images of Figures 2 and 3 clearly show an improvement in surface topography with gas cluster ion doses. With a high-energy dose the scratches were completely eliminated. However, a new high frequency roughness component emerged that resulted in worse or only slightly improved R_a values. This is also seen in the Ar data of Figure 4, where a very low dose results in an increase in R_a followed by a monotonic decline with increasing dose to a roughness level below that of the unprocessed surface. A similar effect was observed by Fathy et al who reported an initial increase in SiC roughness with subsequent smoothing upon increased gas cluster ion dosage [2]. This effect was attributed to defects introduced during wafer polishing based on TEM cross-section analysis. These defects were initially exposed during the first GCIB dose creating the initial roughness and then etched away by the following GCIB dose. A similar effect may be at work here.

A single high Ar dose at low energy was not able to achieve an improvement in surface topography, however it was able to improve a surface that was already treated at a high energy. As seen in Figure 4 the best improvement in surface quality is already achieved at a dose of less than half the maximum dose shown on the plot. At this dose, polishing scratches have been completely removed and higher doses offer only slight improvements in R_a. Improvement seems to be limited by the residual defects in the wafer. A low energy Ar ion dose did not seem to produce as much high frequency roughness and was able to improve the high energy treated surface in a dual energy smoothing process resulting in the lowest surface roughness values achieved with Ar clusters.

A similar effect appears to hold true for polishing with O_2 clusters. Cluster ion doses at high energies are able to remove the mechanical polishing scratches, however they leave a high frequency roughness component similar to processing with Ar. Also as with Ar clusters, O_2 cluster processing at a lower energy removed the high frequency roughness and resulted in a smooth surface. In fact, a dual energy process utilizing oxygen was able to reduce the surface roughness to the lowest level of all the conditions studied in this investigation. The greater reduction of roughness by an O_2 dual energy process over the Ar dual energy process can be clearly seen in Figure 4

The smallest achievable roughness may be limited by the presence of subsurface defects in the SiC induced by the mechanical polishing process. The difference in final roughnesses between the argon and oxygen processes may be the result of fewer defects in the oxygen-processed surface. This seems to be indicated by the RBS ion channeling measurements, which showed a lower number of lattice defects in the oxygen processed surface. This difference may be due to a higher material etch rate for oxygen which allows the exposure of SiC carbide beneath the mechanical damage layer. Etch depth measurements are necessary for further investigation.

Gas cluster ion beam smoothing appears to be a viable process for improving the surface quality of electronic grade SiC. CMP polishing scratches were removed and maximum peak to valley height was decreased overall by over 60% by using a dual energy oxygen cluster ion beam process.

ACKNOWLEDGEMENTS

The authors would like to thank W.R. Brooks and G. Degankolb of the Epion Corporation for their help with the experiments. This work was supported by a fellowship grant from the Engineering Research Program of the Office of Basic Energy Sciences at the Department of Energy and at the Epion Corporation by DoC-NIST ATP (Grant # 70NANB8H4011).

REFERENCES

[1] Bennett, J. A., Holland, O. W., Budde, M., Thomas, D. K. and Feldman, L. C., 2000, "Complete Exfoliation of 4H-SiC by H+ and Si+ Coimplantation," Applied Physics Letters, Vol. 76, pp. 3265-3267.

[2] Fathy, D., Holland, O. W., Liu, R., Wosik, J. and Chu W. K., 2000, "Application of Cluster Ion Beam Smoothing to SiC and YBCO Surfaces," Materials Research Society Symposium Proceedings, Vol. 585, pp. 313-318.

[3] Northby, J. A., Jiang, T., Takaoka, G. H., Yamada, I., Brown, W. L. and Sosnowski, M., 1993, "A Method and Apparatus for Surface Modification by Gas-Cluster Ion Impact", Nuclear Instrumentation and Methods in Physics Research, Vol. B 74, pp. 336-340.

[4] Yamada, I., Matsuo, J., Insepov, Z., Takeuchi, D., Akizuki, M. and Toyoda, N., 1996, "Surface Processing by Gas Cluster Ion Beams at the Atomic (Molecular) Level," Journal Vacuum Science and Technology, Vol. 14A, pp. 781-785.

[5] Yamada, I., 1996, "Applications of Gas Cluster Ion Beams for Materials Processing," Materials Science and Engineering, Vol. A 217/218, pp. 82-88.

[6] Insepov, Z., Yamada, I., and Sosnowski, M., 1997, "Surface Smoothing with Energetic Cluster Beams," Journal of Vacuum Science & Technology, Vol. 15A, pp. 981-984.

[7] Chu, W. K., Li, Y. P., Liu, J. R., Wu, J. Z., Tidrow, S. C., Toyoda, N., Matsuo, J. and Yamada, I., 1998, "Smoothing of YBa2Cu3O7- Films by Ion Cluster Beam Bombardment", Applied Physics Letters, Vol. 72, pp. 246-248.

[8] Fenner, D. B., Torti, R. P., Allen, L. P., Toyoda, N., Kirkpatrick, A. R., Greer, J. A. DiFilippo, V. and Hautala, J., 2000, "Etching and Surface Smoothing with Gas-Cluster Ion Beams," Materials Research Society Symposium Proceedings, Vol. 585, pp. 27-32.

[9] Toyoda, N., Lee, K. K., Luan, H. C., Lim, D. R., Agarwal, A. M., Wada, K. L., Kimerling, C., Allen, L. P. Fenner, D. B., and Kirkpatrick, A. R., 2000, "Surface Smoothing of Polycrystalline Si Waveguides With Gas-Cluster Ion Beams," Materials Research Society Symposium Proceedings, Vol. 597, pp. 51-55.

[10] DiFilippo, V., Saigal, A., Zide, B. M., 2000, "Gas Cluster Ion Beam Polishing of Diamond Machined Optics," Proceedings, 3rd World Conf. on IMPS, pp. 431-437.

[11] Fenner, D. B., Hautala, J., Allen, L. P., Greer, J. A., Skinner, W. J. and Budnick, J. I., 2000, "Smoothing Thin Films with Gas Cluster Ion Beams," Materials Research Society Symposium Proceedings, Vol. 614, pp. F10.3.1-3.6.

[12] DiFilippo, V., Saigal, A., Fenner, D. B., Allen, L. P., Hirvonen, J. K., 2000, "Gas Cluster Ion Beam Smoothing of Silicon Carbide," To be published: Proceedings, THERMEC 2000 Conference.

[13] Fenner, D. B., DiFilippo, V., Bennett, J. A., Tetreault, T. G., Hirvonen, J. K., Feldman, L. C., 2001, "Ion-Beam Nano-Smoothing of Sapphire and Silicon Carbide Surfaces," To be published: Proceedings, SPIE 46th Annual Meeting.

Proceedings of
2001 ASME International Mechanical Engineering Congress and Exposition
November 11–16, 2001, New York, NY
EPP-Vol. 1

IMECE2001/EPP-24715

Equilibrium Elasto-Hydrodynamic Analysis in Wafer Slicing Process Using Wiresaw

Liqun Zhu*and Imin Kao†
Manufacturing Automation Laboratory (MAL)
Department of Mechanical Engineering
State University of New York at Stony Brook

Abstract

An improved elasto-hydrodynamic (EHD) model of wiresaw slicing process is summarized in this paper which incorporates the equation of equilibrium for the motion of the wire. Parametric studies are conducted, which encompass different cases in the wiresaw slicing process. Based on the results, we examine the effects of wire bow angle on the EHD properties and the slicing mechanism of wiresaw process. The simulation results provide insights to the wiresaw slicing mechanism, and suggest ways to facilitate the experimental study for industry wiresaw slicing.

1 Introduction

Wiresaw has been widely used to slice large-diameter silicon wafers (300mm or larger) simultaneously from a crystal ingot, after cropping and trimming. It has gradually become the standard wafer manufacturing technique due to its unique advantages, such as low kerf loss, good wafer surface quality, and high productivity.

As a typical case of three-body abrasion, wiresaw's free abrasive machining (FAM) process involves multi-disciplinary physical phenomena. Among which, the dominant factors are the motion of a slender wire (150 $\sim$ 175μm in diameter) and the tribological behavior of abrasive-carrying slurry. During wiresaw slicing process, the wire, which is commonly made of stainless steel, moves axially at relative high speed (10 to 15m/sec) under high tension (20 to 35N). On the other hand, the wire is also fed at 200 to 500$\mu m/min$ in transverse direction to maintain a proper wire bow (typically several degrees) for cutting, as shown in Figure 1. Meanwhile, hydrodynamic pressure is generated by the moving wire and slurry fluid. The slurry fluid is brought into the cutting region by the wire and/or slurry manifold, and trapped in between the moving wire and ingot surface. By taking the transverse feeding ($-y$ direction in Figure 1) of the wire as quasi-static, the wire would deform until an equilibrium is reached due to this developed hydrodynamic pressure and the wire bow-induced transverse force component of wire tension. Based on the *rolling-indenting* wiresaw cutting mechanism [3, 4, 5, 9], the hydrodynamic pressure and the slurry film thickness under this equilibrium condition are crucial in determining the abrasive motion pattern and the cutting force, which affect the material removal rate and the quality of the sliced surfaces [1, 10]. Typical schematic of modern wiresaw systems can be found in [1, 7].

The development of multi-physics wiresaw slicing model and the investigation of the intertwined physical phenomena remain an important research topic. Up to now, the elasto-hydrodynamic (EHD) model proposed in [1] is the only model through which the basic elasto-hydrodynamic interaction in wiresaw process could be analyzed quantitatively. This EHD model involves coupled Reynold's equation of lubrication and the elastic properties of the compliant wire under tension. The results of the numerical analysis suggest that *floating machining* (indirect contact) environment holds for typical wiresaw slicing process, which gives reasonable explanation to many experimental observations. As a model for initial research however, there exist some drawbacks in this EHD model. For example, it was not comprehensive enough to incorporate the equation of motion of a moving wire, which re-

*Research Assistant; email: zhu@mal.eng.sunysb.edu
†Associate Professor; email: kao@mal.eng.sunysb.edu

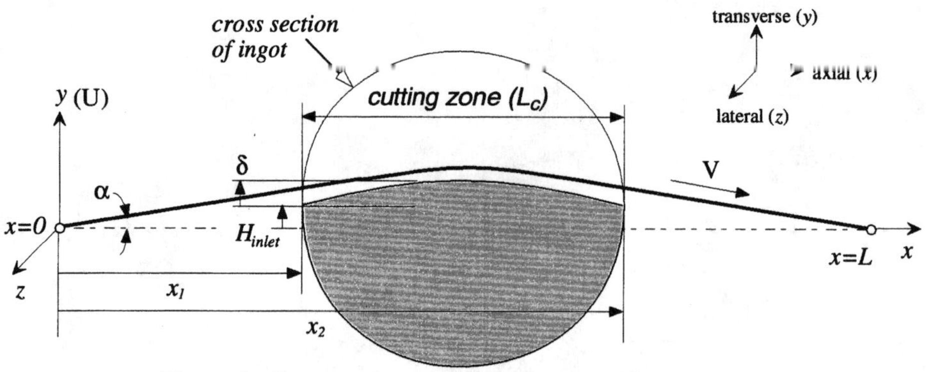

Figure 1: Computation model for wiresaw slicing process

sulted in the inaccurate representation of the wire transverse stiffness.

The objective of this paper is to present our work of an improved model to simulate the EHD interaction in wiresaw slicing process. In this new model, the governing equation describing the equilibrium force-displacement relationship of moving wire is incorporated. Meanwhile, a more efficient approach is employed for the numerical analysis of the coupled nonlinear governing equations. Compared with the previous model, the improved model is found to provide better accuracy and to improve computational efficiency.

2 Computation model

The computation model of wiresaw slicing is shown in Figure 1. The taut wire travels axially at speed V over the substrate surface. From experimental results, the wire marks representing the cutting profiles along the axial direction of the wire is found to match approximately with a segment of circular arc of very large radius of curvature. This wire mark curve on the ingot is denoted by $\delta(x)$. The bow angle of the wire, α, is mainly determined by H_{inlet}, which is the distance that the slurry entry point projects into the horizontal line between the two wire guides. The parameter x_1 denotes the x coordinate of the beginning of the cutting zone. Once the slurry fluid enters the converging passage between the wire and the ingot, elasto-hydrodynamic effect occurs [1].

Due to the fact that the transverse motion of wire caused by vibration is dominated by its tension [7], the following assumptions are reasonable to this EHD model and corresponding analysis:

1. The wire has negligible flexural rigidity;
2. Thin film formulation is assumed in the analysis so that there is no variation of pressure along the film thickness;
3. The slurry flow behaves as a Newtonian fluid [1]; and
4. Wire tension, translating speed and mass density (per unit length) are uniform and constant.

The equilibrium displacement of the wire satisfies [7]:

$$(m V^2 - T) U_{xx} = f_w [\mathcal{H}(x - x_1) - \mathcal{H}(x - x_2)] \quad (1)$$

where m is the mass density of the wire per unit length, U is wire transverse displacement, V is the axial moving speed of the wire, T is the wire tension, x_2 is the position of the exit point, and $\mathcal{H}$ stands for Heaviside function. The subscript "x" denotes the partial derivatives with respect to x (Figure 1), for example U_{xx}, and f_w is the distributed force acting transversely on the wire per unit length, which can be obtained by integrating the pressure acting on the wire along wire circumference. The boundary conditions associated with equation (1) are:

$$U(0) = U(L) = 0 \quad (2)$$

Under equilibrium condition, the flow of slurry follows the steady-state 2-D Reynold's equation of lubrication for incompressible fluid [6]:

$$(h^3 p_x)_x + (h^3 p_y)_y = 6\mu V h_x \quad (3)$$

where h is the slurry film thickness, p is the EHD pressure in the slurry, and μ is the dynamic viscosity of the slurry flow. Subscript stands for partial derivative. Swift-Steiber boundary conditions is applied to equation (3) on the lubrication boundaries:

$$p = 0 \quad (4)$$

$$\frac{\partial p}{\partial n} = 0 \quad (5)$$

where n is the outward normal to the boundary. The 2-D slurry film thickness h is composed of two parts: the geometry-dependent film thickness h_g and the minimum film thickness h_0, as shown in Figure 2.

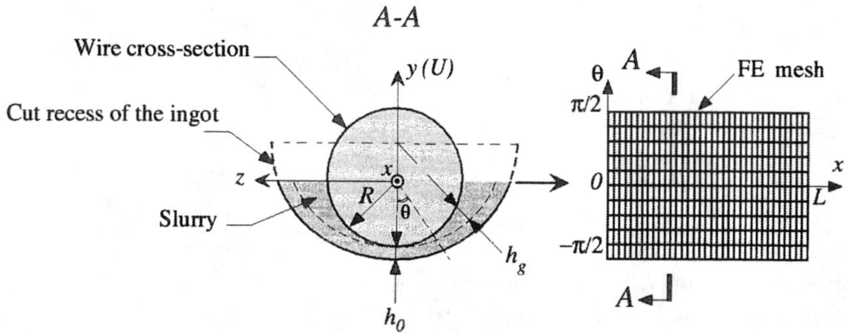

Figure 2: Slurry film thickness and the FE mesh

3 Numerical Scheme

The set of governing equations discussed in Section 2 are characterized by their coupling physics and nonlinearity. Moreover, their dimensional incompatibility raises difficulties in the corresponding numerical analysis. To solve these problems, the numerical scheme of incorporating FEM and nonlinear inexact Newton-GMRES iteration are adopted in this equilibrium EHD interaction analysis.

At equilibrium state, the magnitude of the discretized error, **e**, in the following equation (6) should be minimized after FE discretization

$$\mathbf{e}(\mathbf{h_0}) = \mathbf{h_0} - [\mathbf{U} - (\boldsymbol{\delta} + H_{inlet}\mathbf{a})] = 0 \quad (6)$$

In equation (6), let n be the nodal degree of freedom of the wire after discretization, then $\mathbf{h_0} \in \mathcal{R}^n$ and $\mathbf{a} \in \mathcal{R}^n$, $a_i = 1\ \forall i \in [1, n]$, and $\mathbf{e} : \mathcal{R}^n \Rightarrow \mathcal{R}^n$ is the nonlinear mapping. By solving this equation, we obtain the equilibrium minimum film thickness $\mathbf{h}_0^*$. The *Inexact Newton-GMRES* iteration method is employed in the corresponding numerical analysis to improve computation efficiency. The flowchart of the numerical analysis is illustrated in Figure 3.

4 Results and Discussion

Parametric studies are conducted using this improved EHD model proposed in the previous sections. Typical wire parameters for an industrial wiresaw process are used in the simulation.

1. **Total wire span between supporting wire guides:** $L = 500mm$ (symmetrically span over the ingot);

2. **Contact span between wire and ingot:** varies from 0 to $100mm$;

3. **Mass density of stainless steel wire:** $m = 0.1836g/m$; and

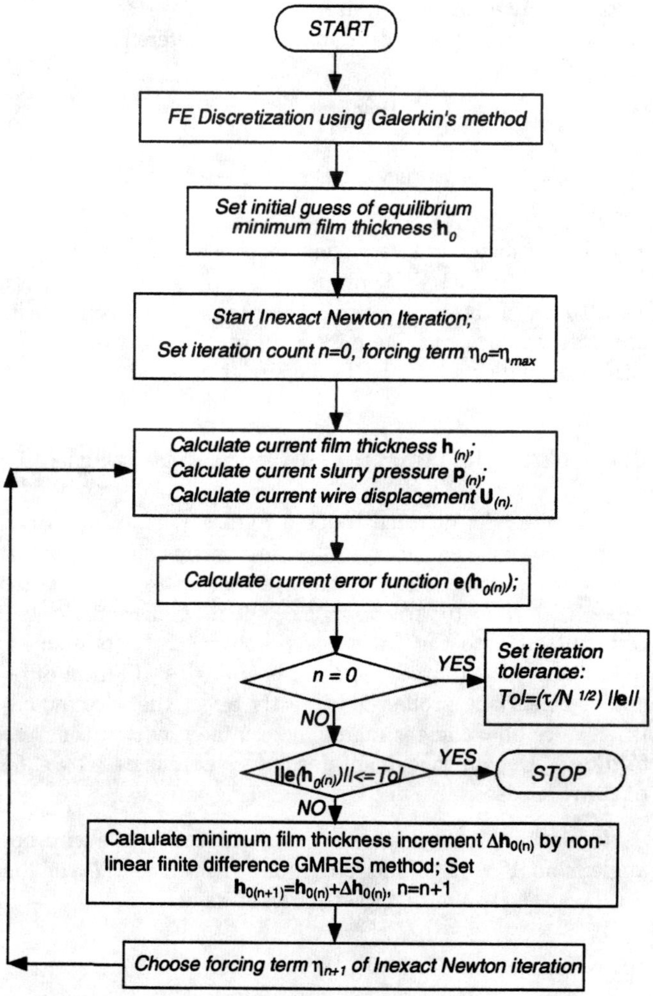

Figure 3: Flowchart of the numerical analysis

4. **Radius of wire:** $R = 87.5\mu m$.

Experimental observations show that the morphology of wiresaw-sliced wafer surface has a random wear pattern, which consists of a series of closely spaced pits and indentations. There is little sign of any directionality to the surface deformation of the ingot surface. This is typical surface morphology produced by wet FAM process when the film thickness is larger than abrasive particle size [1, 8]. Also, from our previous simulation results, stable floating machining[1] condition usually can be achieved during wiresaw slicing with a set of well-chosen process parameters. Under floating machining condition, the abrasives remove the substrate material by typical "rolling-indenting" mechanism. With constant slurry viscosity and abrasive grit size, the material removal rate R_{mr} is directly related to hydrodynamic parameters [1]

$$R_{mr} \propto V^2 \frac{p_{avg}}{h_{avg}^2} \tag{7}$$

where p_{avg} is the average EHD pressure and h_{avg} is the average film thickness.

Wiresaw slicing is a very complex process. There are many factors affecting its EHD properties, such as bow angle, tension, speed of the wire, slurry concentration, slurry viscosity, abrasive grit size, etc. Here we focus on the effect of different wire bow angles on the EHD properties of wiresaw slicing.

4.1 Parametric study of wire bow angle

Wire bow angle, denoted by α in Figure 1, is an important parameter that can be controlled in wiresaw slicing process. The vertical component of the wire tension results in transverse force due to wire bow angle. This transverse force is transmitted onto the abrasives via slurry EHD pressure to provide indenting force. Different bow angles will form different ingot surface profile and directly affect the wiresaw cutting. Wire bow angle is mainly determined by the transverse feed rate. Usually it is maintained at several degrees ($\alpha \leq 6°$) to provide reasonable and stable cutting force.

Simulation has been performed by increasing the wire bow angle from $1° \sim 2°$. The length of cutting zone (L_c in Figure 1) is set at $50mm$. Other parameters are

1. Wire tension: $T = 20N$, wire speed: $V = 15m/sec$;
2. Slurry viscosity: $\mu = 5000cP$ (F400 Green/Ecocool slurry with 1.0 Kg/liter grit/base at room temperature [2])

[1]Floating machining happens when there is no direct contact between wire, abrasive and ingot surface.

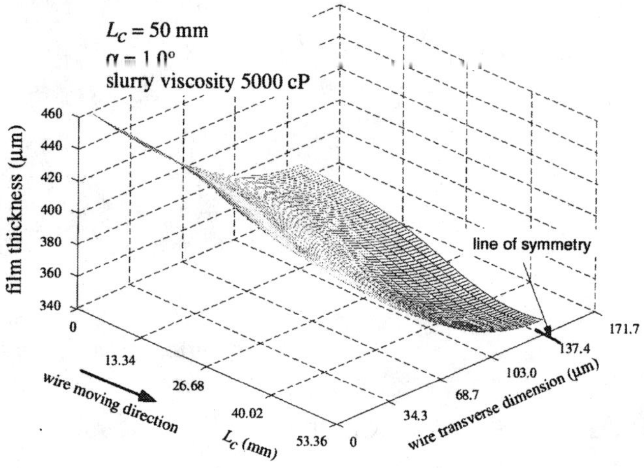

Figure 4: The profile of film thickness at $\alpha = 1°$

The simulation results of the EHD profiles at 1° and 1.75° are shown in Figure 4, 5, 6 and 7. The parameter $\frac{p_{avg}}{h_{avg}^2}$ in equation (7), together with the rate of the film thickness variation $\frac{\Delta h_{0min}}{L_{eff}}$, are shown in Figure 8 with different wire bow angles[2].

From the EHD profiles, we found convergent channel covering most part (or all) of the ingot surface always can be formed for the slurry flow, which indicates that stable floating machining condition for wiresaw process can be achieved. Twin-peak pattern is the typical feature of the EHD pressure profiles. The higher peak appears at the position right after the middle of the ingot surface. This is because of the rapid change of the ingot surface profile and relatively small film thickness at the middle of the ingot surface. The lower peak is very important in opening the slurry flow passage at the beginning part of the cutting zone and in forming a continuously convergent channel to reach maximum EHD load capacity. The area covered by the higher peak is much larger than that covered by the lower one – indicating that the cutting action is dominated by the EHD pressure within the higher peak portion. In the simulation of slicing circular ingot, this "twin peak" pattern is very typical in EHD pressure profiles, while the magnitudes and positions of two peaks may vary at different configurations (slurry viscosity, ingot surface curvature, etc.)

4.2 Discussions

The highlights of Figure 8 together with the EHD profiles are listed in the following.

[2]L_{eff} is the length of ingot surface covered by a positive pressure.

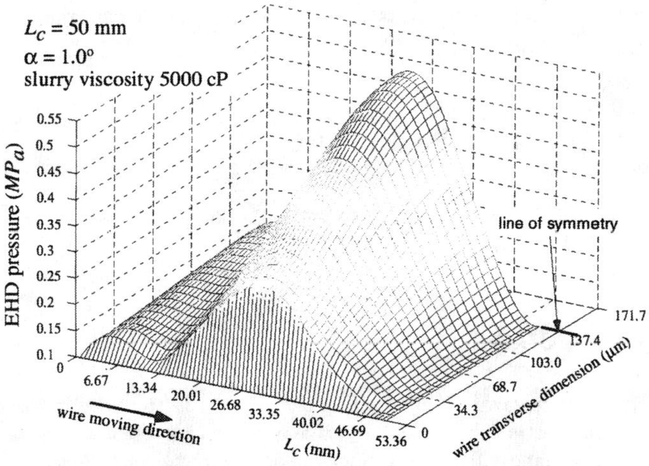

Figure 5: The EHD pressure profile at $\alpha = 1°$

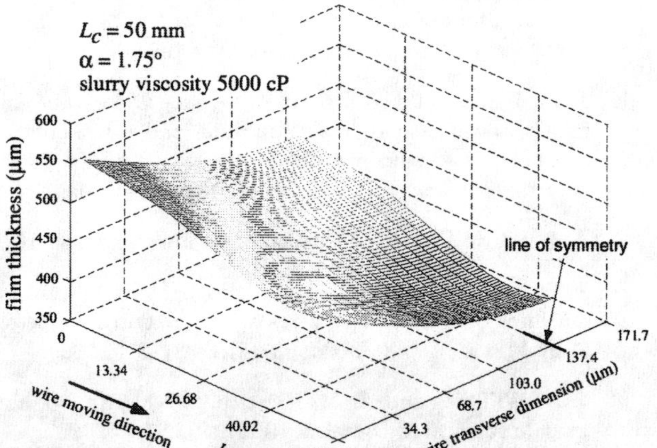

Figure 6: The profile of film thickness at $\alpha = 1.75°$

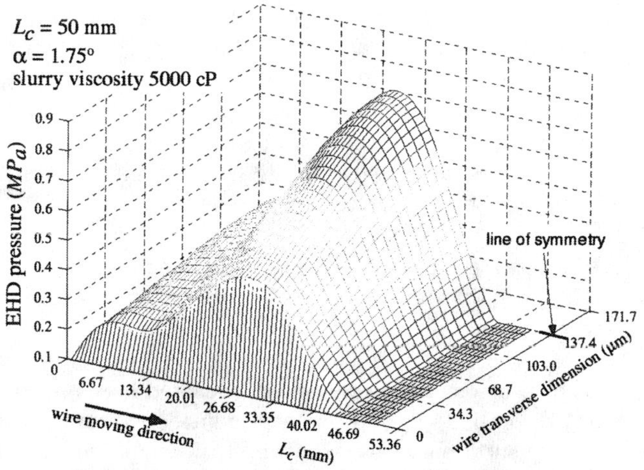

Figure 7: The EHD pressure profile at $\alpha = 1.75°$

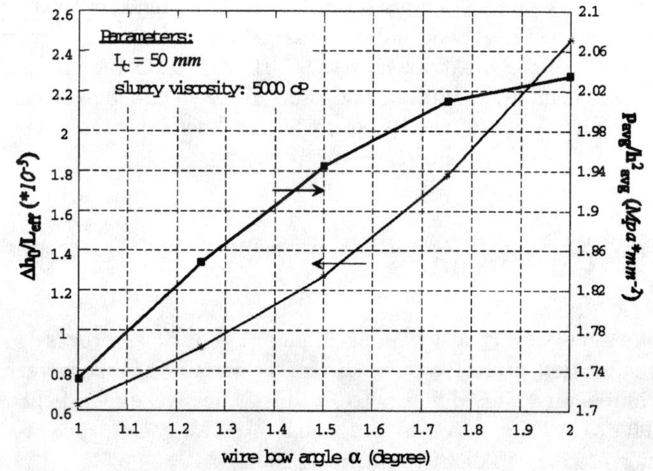

Figure 8: The rate of film thickness variation and parameter $\frac{p_{avg}}{h^2_{avg}}$ vs. the bow angle α

1. When wire bow angle α is larger, both the rate of film thickness variation and the parameter $\frac{p_{avg}}{h_{avg}^2}$ increase. First, this indicates that the EHD load capacity is enhanced mainly though forming larger rate of film thickness variation when the bow angle becomes larger. Secondly, the material removal rate R_{mr} also increases with larger wire bow angle based on equation (7).

2. In wiresaw slicing, the kerf loss should be minimized. By increasing wire bow angle, faster cutting can be achieved with a compromise of increased kerf loss.

3. At a specific cutting zone length (denoted by L_c in Figure 1), both H_{inlet} and the longitudinal curvature of the cutting surface increase with wire bow angle. Under this circumstance, the pressure build-up in the slurry becomes inadequate in opening a convergent slurry channel covering the whole cutting zone, as shown in Figure 6 and 7. Under floating machining condition, the abrasive particles tend to take rolling motion and cannot cause any machining in the absence of positive hydrodynamic pressure. Therefore, large bow angle results in the shrinkage of effective cutting length (the part of substrate surface covered by a positive pressure to facilitate the rolling-indenting process) along ingot surface. Furthermore, the length decrement of the convergent slurry passage becomes more and more serious as the bow angle continuous to increase. Eventually, too large a wire bow angle may cause the breakdown of the EHD condition for floating machining condition, and cause direct contact to occur. This is one of the reasons why the bow angles are typically kept low in a wiresawing process.

5 Conclusions

Wiresaw slicing is becoming standard industrial process in slicing thin crystal wafers. However, wiresaw slicing is very complicated and difficult to conduct effective experiments. Therefore, simulation is important in investigating the interior mechanism of wiresaw process.

In this paper, a more comprehensive elasto-hydrodynamic model for wiresaw slicing process is presented which involves the equilibrium wire motion. Comparing to our previous EHD model, this improved model adopts more accurate representation of the moving wire and its transverse stiffness. When applying this EHD model to typical wiresaw process, numerical simulation is performed to investigate EHD properties under different wire bow angles. Twin-peak pattern is observed in the EHD pressure profiles under floating machining condition. Simulation results also indicate that too large a wire bow angle may cause the breakdown of floating machining condition.

6 Acknowledgment

This research has been supported by the NSF grants: DMI-9634889 and DMI-0085021.

References

[1] M. Bhagavat, V. Prasad, and I. Kao. Elasto-hydrodynamic interaction in the free abrasive wafer slicing using a wiresaw: Modeling and finite element analysis. *the ASME Journal of Tribology*, 122(2):394–404, April 2000.

[2] HCT Corp. Annual report for wiresaw project. Technical report, GT Equipment Technologies, Inc., March 1997.

[3] I. Kao, M. Bhagavat, and V. Prasad. Integrated modeling of wiresaw in wafer slicing. In *NSF Design and Manufacturing Grantees Conference*, pages 425–426, Monterey, Mexico, January 5-8 1998.

[4] I. Kao, V. Prasad, F. P. Chiang, M. Bhagavat, S. Wei, M. Chandra, M. Costantini, P. Leyvraz, J. Talbott, and K. Gupta. Modeling and experiments on wiresaw for large silicon wafer manufacturing. In *the 8th Int. Symp. on Silicon Mat. Sci. and Tech.*, San Diego, May 1998.

[5] J. Li, I. Kao, and V. Prasad. Modeling stresses of contacts in wiresaw slicing of polycrystalline and crystalline ingots: Application to silicon wafer production. *ASME Journal of Electronics Packaging*, 120(2):123–128, June 1998.

[6] A. Z. Szeri. *Tribology*. Hemisphere Publishing Co., 1979.

[7] S. Wei and I. Kao. Vibration analysis of wire in free abrasive machining of the modern wiresaw manufacturing process. *Int. J. of Vibration and Sound*, 231(5):1383–1395, 2000.

[8] J. A. Williams and A. M. Hyncica. Mechanisms of abrasive wear in lubricated contacts. *Wear*, 152:57–74, 1992.

[9] F. Yang and I. Kao. Interior stress for axisymmetric abrasive indentation in the free abrasive machining process: Slicing silicon wafers with modern wiresaw. *ASME Journal of Electronics Packaging*, 121(3):191–195, September 1999.

[10] L. Zhu, M. Bhagavat, and I. Kao. Analysis of the interaction between thin-film fluid hydrodynamics and wire vibration in wafer manufacturing using wiresaw. In *the Proc. of IMECE'00: EEP-Vol 28* packaging of electronic and photonic devices, pages 233–241, Florida, Nov. 2000. ASME Press.

Proceedings of
2001 ASME International Mechanical Engineering Congress and Exposition
November 11–16, 2001, New York, NY
EPP-Vol. 1

IMECE2001/EPP-24716

A Neural-Taguchi Based Quasi Time-Optimal Control Strategy For Chemical-Mechanical Polishing Processes

Gou-Jen Wang and Meng-Hsian Chou
Department of Mechanical Engineering
National Chung-Hsing University
Taichung 402, Taiwan

ABSTRACT

Besides the major factors such as the down force, back pressure and the rotating speed of wafer carrier, effect of polishing time is also an important issue in CMP processes. In this study, a neural-Taguchi method based cost-effectively quasi time-optimal technique for Chemical-Mechanical Polishing (CMP) processes is developed. Key concept of this new technique is that an optimal processes parameter set is obtained through a neural networks simulated CMP processes model. Under such an optimal parameters set, the desired material removal rate (MRR) and within-wafer-nonuniformity (WIWNU) can be reached with the optimal polishing time. It has been proved by experiments that the proposed method can offer a better polishing performance while reducing the polishing time by 1/3.

INTRODUCTION

Chemical mechanical polishing (CMP) is the process of smoothing and planing and is aided by chemical etching and mechanical grinding forces. CMP is achieved by bringing the oscillating wafer surface into contact with a rotating polishing pad as the tiny particles contained slurry is applied to the wafer/pad interface.

Planarization of wafer is the primary goal that CMP techniques intend to achieve. Both the resolution level of deposition layer and the completeness of pattern movement during photo-lithography process are affected by planarization. When the width of the conductor in an ultra large-scale integrated circuit is narrower than 0.25 μm, CMP is the only technique that could provide a global planarization [1][2]. Because of this, CMP process technique has drawn attention from the semiconductor industry and becomes a key technique being developed by major factories and research institutions.

Because CMP processes include the effects of both chemical reaction and mechanical grinding, its action fundamentals and models are still not quite clear. Also, many factors during the process [3], such as rotating speed of wafer carrier, pressure on the wafer, the supply rate of slurry and the PH of slurry etc., have great impacts on polishing quality. How to precisely control the processes and get appropriate removal in polishing avoiding excessive or insufficient removal is another important research subject of CMP processes, which is called end point detection. Traditional methods of end point detection are applied off-line. The off-line measurement is accurate but time-consuming, which causes excessive or insufficient removal. Currently, most of the developing methods are toward the direction of on-line measurement.

End point detection can be divided into two main categories: removal of the same material and removal of different materials. Therefore, among the reported end point detection techniques, there are also two categories: frictional force change related and not related. The fundamental idea of end point detection methods based on the concept of frictional force is that removal of different materials has different frictional forces. The major methods are motor current method [4] and pad thermal image method [5]. Motor current method detects the current change of carrier motor and decides whether to stop the removal action of CMP processes. The engagement of different materials during polishing processes will change the frictional force and the measured motor current will change accordingly. This change claims the end point of CMP process. This method does not need to change the machinery structure and is the simplest to apply except that it is only applicable to removal of oxidized layer. Pad thermal image method detects the end point by the temperature change of polishing pad. The fundamental theory is that the changes in temperature are caused by chemical reaction from slurry during CMP process and fluctuations of frictional force with the engagement of different materials. So the measurement of pad temperature could determine the end point of polishing. The tool used in this method is the IR thermal sensor installed in the pad. Because of the interference of surrounding environment, this method is not accurate enough. As to the removal of same material, one of the representative methods is optic interferometry method [6][7].

The disadvantage of this method is that the original structure of CMP machine must be modified. The acoustic method detects the thickness variation of the polishing layer through the change of acoustic amplitude and frequency. Small vibration method could measure the change of vibration while the wafer is approaching planarization. The conductivity and impedance methods detect the end point through the change of wafer conductivity and impedance, which are only practical to detect the metal layer removal.

Besides production, planarization and surface uniformity of the wafer, the cost of ownership (COO) is another important subject. This goal could only be achieved through the combination of accurate on-line monitor, end point detection technique and precise closed-loop process control technique. However, a real instant closed-loop process control strategy has not been proposed to date. How to combine accurate on-line monitoring and end point detection techniques with the development of an instant closed-loop control technique to increase the production and reduce COO is an attractive topic deserves to be worked on.

In this study, based on the existing CMP process parameter optimization techniques [8] and the mechanical wear mechanism of CMP [9], with the consideration of polishing time effect on removal rate and planarization, we hope to develop a new CMP end point detection technique which doesn't require modification of current machine structure or any additional instrument. In another word, this new technique is able to end the polishing processes through polishing time control. To get a high removal rate and the minimum WIWNU, we use neural Taguchi method [10] to design the experiment and construct the process model.

NEURAL-TAGUCHI METHOD

Neural Taguchi experimental method is a combination of neural networks and Taguchi experimental method. The effectiveness has been provided in the parameter design applications [11][12]. The followings are the steps of CMP process parameter optimization by neural Taguchi method:

Step1. Use Taguchi method to design and conduct the experiments.

Step2. Use neural networks to model the relationship between CMP process parameters and MRR & WIWNU, respectively.

Step3. Use the signal to noise ratio analysis (S/N analysis) to obtain the local optimal process parameters for MRR and WIWNU.

Step4. Fine-tune resulting parameters by using the neural-network constructed in step 2.

Step5. Verify the results by additional experiment.

Design of Experiment

Processes Parameters

The experimental materials in this study are mainly oxidized layers, which are easy to obtain. According to the results from references [8] and [9], we chose four important and adjustable process parameters as control variables. They are Down Force, Platen Speed, Solid Content, and Back Pressure. The relationship between removal rate and polishing pressure & polishing speed can be described by Preston equation [13]:

$$Removal\ \ Rate = \frac{\Delta H}{\Delta t} = k_P P \frac{\Delta s}{\Delta t},$$

where ΔH is the change of surface height, Δt is the polishing time, P is polishing pressure, $\Delta s/\Delta t$ is the linear relative speed of pad to wafer, and k_p is the polishing parameter. We can see that the removal rate is linear dependant of the polishing pressure and the polishing time. i.e. the higher the polishing pressure, the faster the removal rate. In Preston equation, the polishing pressure is equal to the back pressure divided by the contact area of polishing pad. Wafer surface uniformity will change with the polishing time, so will the contact area. As a result, the back pressure will change accordingly. We can predict that the MRR and WIWNU are also functions of the polishing time. From experimental data of Fig. 1 and Fig. 2, we conclude that MRR and WIWNU definitely change with time. Therefore, polishing time has been chosen as the fifth control variable to construct a more comprehensive CMP process model.

Measurement Variables

In the design of experiment, an L_{25} (5×5) orthogonal array [14] as shown in Table 1 is adopted. This orthogonal array has 25 columns 5 rows, where 5 rows represent 5 control variables from A to E. Five levels on each variable covering the domain of interest are set. The dependent variables to be observed are MRR and WIWNU, where MRR is the bigger the better and WIWNU is the smaller the better. The procedures for measuring MRR and WIWNU are briefly described as below.

(1). Use the AFT model 2100UV by Nanospec to measure the thickness of wafer before polishing (nine points as shown in Fig. 3). Set it as t_1

(2). After T minutes, measure the thickness of wafer after polishing. Set it as t_2.

(3). Polishing rate = $(t_1 - t_2) / T$. MRR = the average polishing rate of all the points.

(4). WIWNU = (standard deviation of t_2)/(mean of t_2) × 100%

Since it only makes sense to compare WIWNUs after wafers have been polished and have similar thickness, we replace WIWNU with standard deviation of t_2 in this study. In the follow-up experiments, we will use experimental data to demonstrate that WIWNU can be replaced by standard deviation of t_2, and find the optimal parameters in accordance with WIWNU.

The experiments have been done at the National Nano Device Laboratories (NDL) of Taiwan. The CMP machine used is the Westech 372. The diameter of the polishing platen is 57 cm. The wafers used in our experiments are silicon wafer of 15.24 cm (6 inch) diameter. The slurry used is the Cabot SS25 (fume silica 12 wt. %, KOH formulated at pH 10.5).

Experimental results and their relative S/N ratio are illustrated in Table 2.

Signal/Noise Ratio Analysis

Basically, control variables and noises are major factors affecting process performance. Taguchi experimental method takes signal/noise ratio (abbreviated as S/N ratio) as the performance index to evaluate the quality of product or process. Statistically, S/N ratio is the ratio of mean of measurements to variance of measurements. A higher S/N ratio means a more consistent quality and a better system performance.

As to the CMP process, after completing orthogonal array experiments, S/N ratio of MRR and standard deviation of t_2 (σ_{t2}) in each experiment are calculated first, then the effect of each factor is determined and corresponding responding plots and tables are constructed, finally ANOVA analysis is executed. Responding tables show the effect of each factor at different levels and responding plots illustrate the trend of effect changes between levels. ANOVA analysis describes the contribution of each factor (the effect of each factor). From the responding tables and responding plots, we can find the value of each factor that can maximize S/N ratio, i.e. the optimal value. Combining all the factors at the optimal level, we get the optimal parameter set with regard to MRR and WIWNU (in terms of σ_{t2}).

For the experimental results in Table 2, the corresponding responding tables and responding plots are shown in Table 3, Table 4, Figure 4, and Figure 5. From Figure 4, we can see that $A_5B_5C_1D_5E_1$ is a local optimal parameter set for MRR. And from Figure 5, we see that $A_3B_1C_4D_2E_1$ is a local optimal parameter set for WIWNU.

The key function of ANOVA is to decide the relative importance of different factors. ANOVA partitions the variations of data into squared sum of effect of factors, correlative interactions, and errors. Then it investigates the significance of the impact that the factors' major effects and their correlations have on the variance of error. In this study, ANOVA analysis will help us to find the contribution of each factor to increasing MRR or decreasing WIWNU within the preset range, which is a good reference for the follow-up study.

ANOVA analysis in this study is done with reference to relevant theorems about Taguchi method [15]. The calculation results are shown in Table 5 and Table 6.

From the results, we can see that the effect of back pressure and polishing time on removal rate is relatively small compared to other factors. But the WIWNU is influenced by all five control variables to a certain extent.

CMP Processes Modeling

Neural Network Training Samples

Since CMP is a very complicated system, this study tries to use highly non-linear neural network to simulate it. But in most cases, the relatively simple orthogonal array cannot provide sufficient training samples for neural networks to formulate the completely implicit model. As the result, we cannot accurately obtain the optimal parameters set. It is desired to investigate whether the optimal parameters satisfy the expectation. If not, we shall redo the orthogonal array experiment using parameters values close to the optimal set obtained by Taguchi method. Recursively, the above steps are executed until the desired optimal parameters set is reached.
In this study, we use the experimental results in Table 2 as training samples, and normalize the desired outputs between –0.8 and 0.8 for convenience of neural network training.

Model Construction

Model construction is to simulate CMP process by neural networks. The network used to model CMP processes is depicted in Figure 6. Learning rule for the neural network is the CSDL algorithm proposed in reference [16]. Its advantage is that no human interaction is required to adjust the learning parameters during training period.

In the network training, theoretically we should take those control variables in table 1 as network inputs, and removal rate and σ_{t2} as two outputs. Since we also need to take into consideration the polishing time effect on MRR and σ_{t2} and MRR is the bigger the better and σ_{t2} is the smaller the better, instead, output of the network is chosen to be the hybrid performance index defined as,

$$J = a \times \mathrm{MRR(S/N)} + b \times \sigma_{t2}\ \mathrm{(S/N)}$$

Where (MRR(S/N) and WIWNU (S/N) denote S/N ratio of MRR and σ_{t2} respectively. No matter the bigger the better or the smaller the better, the S/N ratios of MRR and σ_{t2} are the bigger the better when a global optimal parameters set for both MRR and σ_{t2} is desired. It implies that the goal of parameters adjusting is to maximize the performance index *J*. Because the main consideration of CMP processes is fast material removal with WIWNU under the required range, *a* and *b* in *J* are chosen to be 0.6 and 0.4, respectively.

Processes Parameters Optimizing

Since the values at different levels in orthogonal array are discrete, the optimal parameters set from S/N ratio analysis is still a relatively rough estimation. Additionally, the output of the network as shown in Figure 6 is the MRR and σ_{t2} combined performance index. Therefore, the optimal parameters set cannot be obtained from S/N analysis directly. In order to acquire a more precise optimal parameters set, the neural network based implicit CMP model built in Section 2.4 together with the parameters optimization method [11] is adopted to fine-tune the parameters. The fine-tuning steps are as follows:

Step 1: Put the local optimal parameters set from S/N ratio analysis into neural network and calculate network output value $(J)_{max}$.

Step 2: Calculate $\partial(J)/\partial x_i$ around $(J)_{max}$, change x_i within allowable range to increase $\partial(J)/\partial x_i$, if $\partial(J)/\partial x_i$ is positive, increase x_i ; if $\partial(J)/\partial x_i$ is negative, decrease x_i . When x_i is not changeable, jump to step 4. (x_i represents the network input).

Step 3: Use neural network to calculate *J*, if $J > J_{max}$, Set J_{max} = *J* and go back to step 2.

Step 4: The parameters set obtained from the above steps are the optimal parameters with regard to MRR and σ_{t2}.

Following the above steps, the optimal parameters set can be found as:

Solid content: 20%, Down force: 8.0 psi, Back pressure: 3.2 psi, Platen speed: 41 rpm,Polishing time: 42s

The physical meaning of 42 seconds is that when other parameters are set as above, 42 sec is the optimal polishing time. It means that the polishing processes can be stop after 42 seconds to achieve maximal removal rate and acceptable WIWNU.

EXPERIMENTAL VERIFICATION

Experimental verification is to carry out the experiment with optimal parameters set obtained from neural-Taguchi method. For comparison, experiments with parameter settings currently used at the NDL of Taiwan are executed. Experimental results are tabulated in Table 7. Since the carrier speed and oscillation speed have little influence on MRR and WIWNU., the primitive settings of these two parameters by NDL are adopted.

From the results shown in table 7, we find that the approach proposed by this study reduces polishing time substantially to achieve fast material removal with WIWNU within acceptable range. Repeatedly experiments give similar results. It is concluded that 42 seconds is the optimal polishing time for the oxidized layer material used, if the optimal parameters set obtained by the proposed method is implemented. As to other non-oxidized layer materials, we can also find the optimal polishing time through the same process. In another words, the method proposed in this study can systematically choose optimal parameters set in CMP process and be used as an effective tool in CMP end point detection without modifying any machinery structure.

CONCLUSIONS

In this study, based on neural Taguchi method, the number of experiments is effectively reduced without affecting the completeness. In selection of control variables, the polishing time is added for better simulation of the polishing processes. A quasi-optimal parameters set including polishing time was acquired by neural networks based optimizing approach. When this optimal parameters set is implemented, the polishing processes could be completed in optimal time with WIWNU staying in acceptable quality. In the experiment for verification, the proposed method offer a better polishing performance while reducing the polishing time by 1/3.

Additionally, the end point detection technique proposed in this paper does not require any modification of machine structure or additional instrument, and can be applied to wafer of different materials. From the viewpoint of cost or effectiveness, it is highly feasible and profitable.

Acknowledgement

The authors would like to thank the National Science Council of Taiwan for financial support under the grant NSC-89-2212-E-005-018. The authors would also like to thank the National Nano Device Laboratories of Taiwan for providing equipment and materials for the completion of the experiments.

REFERENCES

[1] O'Mara, W. C. "Planarization by CMP: Forecasting the Future", *Semiconductor International*, pp. 140-146, July 1994.

[2] The National Technology Roadmap for Semiconductors, Semi- conductor Industry Association, San Jose, CA, 1994.

[3] Steigerwald, J. M., Murarka, S. P., and Gutmann, R. J., "Chemical Mechanical Planarization of Microelectronic Materials", JOHN WILEY & SONS , INC. 1997

[4] Tsai, M. S., "Short course for the Integrated-Circuit Process", National Nano Device Lab., Taiwan ,1998

[5] Chen, L. J., Huang, Y. L., Lin, Z. H., and Chiou, H. W., "Pad Thermal Image Endpointing for CMP Process", *CMP-MIC Conference* , pp.28-35, Feb 19-20,1998

[6] Chan, D. A., Swedek, B., Wiswesser, A., and Birang, M., "Process Control and Monitoring with Laser Interferometry Based Endpoint Detection in Chemical Mechancial Planarization", *IEEE/SEMI Advanced Semiconductor Manufacturing Conference*, pp.377-384, 1998

[7] Fang, S. J., Barda, A., Janecko, T., Little, W., Outley, D., Hemple, G., Joshi, S., Morrison, B., Shinn, G. B., and Birang, M., "Control of Dielectric Chemical Mechancial Polishing Using an Interferometry Based Endpoint Sensor", *Interconnect Technology Conference*, pp.76-78, 1998

[8] Wang, G. J., Chen, J. L., and Hwang, J. Y., "A New Optimization Strategy For Chemical Mechanical Polishing Process," Accepted for publication in *the JSME International Journal*, 2001.

[9] Tsai, M. Y., Tsai, J. C., and Liu, C., "Mechanical Wear Models and Experimental Investigation for CMP Process," Proceeding of The Sixth Symposium on Nano Device Technology, pp. 20-23, Hsin-Chu, Taiwan, 1999.

[10] Wang, G. J., Tsai, J. C., Tzeng, P. C., "Neural-Taguchi Method for Robust Design Analysis", *Journal of the Chinese Society of Mechanical Engineers* Vol. 19, NO2. pp.223-230.1998

[11] Wang, G. J., Wu, J. G., Chang K., 2000, "A New Neural Network Based Parameters Optimization Method," *Proceeding of The Sixth International Conference on Automation Technology*, Taipei, pp 609-616.

[12] Wang, G. J., Chen, J. L., and Hwang, J. Y., , "A New

Optimization Strategy For Chemical Mechanical Polishing Process," Accepted for publication at *the JSME International Journal*, 2000.

[13] Preston, F., "The Theory and Design of Plate Glass Polishing Machines," J. Soc. Technol., 11, pp. 214, 1927.

[14] Taguchi, G. and Wu, Y. I. "Introduction to Off-Line Quality Control" *Japan Quality Control Association*, 1979

[15] Ross, P., *Taguchi Techniques for Quality Engineering*, McGraw-Hill, New York, 1988.

[16] Wang, G. J., and Chen, J. J., 2000, "Cascade Steepest Descendant Learning Algorithm for Multilayer Feedforward Neural Network," *JSME International Journal*, Series C, Vol.43, No. 2, pp 350-358.

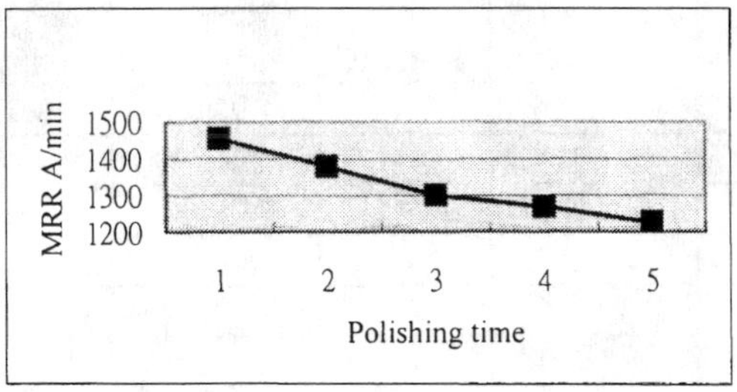

Figure 1. Removal rate vs. polishing time

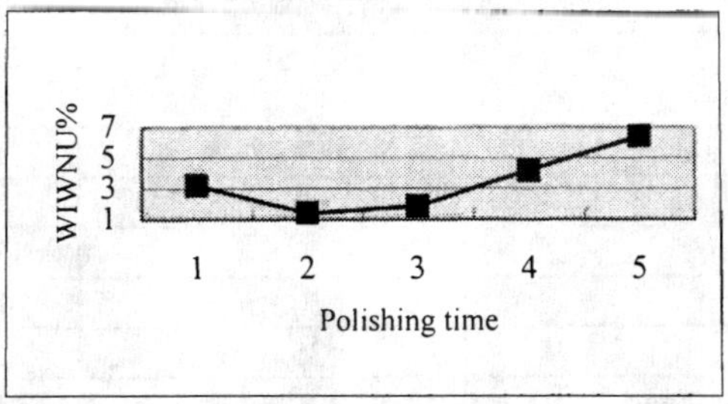

Figure 2. Within-wafer-nonuniformity vs. polishing time

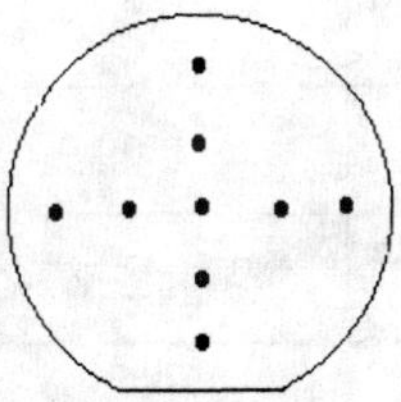

9-point

Figure 3. Distribution of measuring points

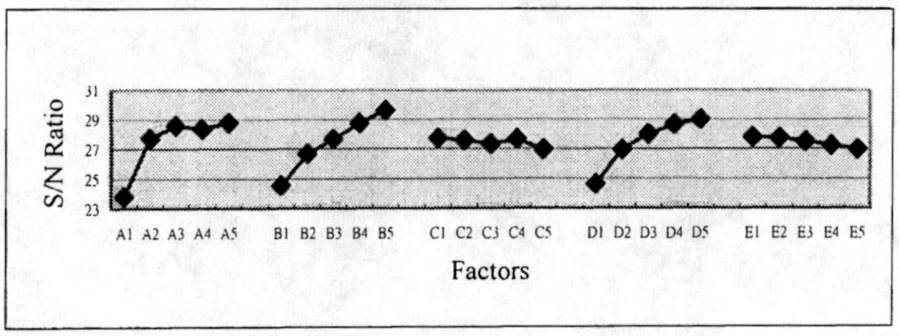

Figure 4. Figure of response for MRR

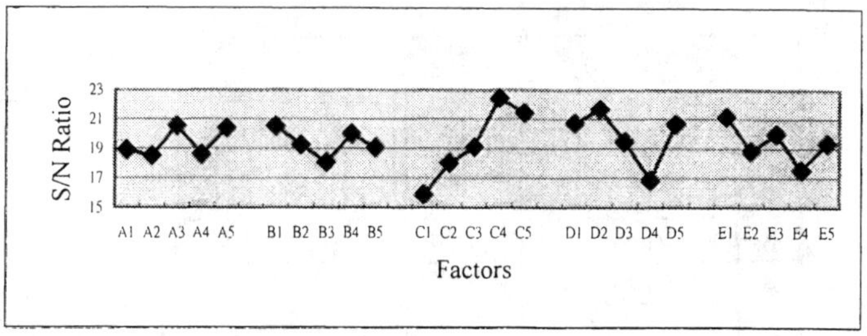

Figure 5. Figure of response for σ_{t2}

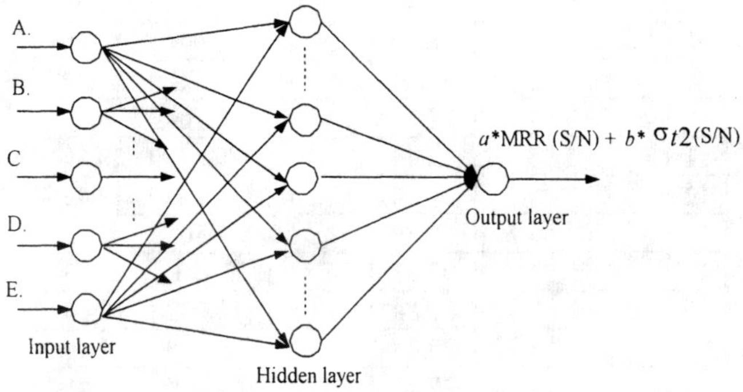

Figure 6. Neural-network configuration for CMP processes modeling

Table 1. $L_{25}(5^5)$ Matrix experiment

Col / Row	$L_{25}(5^5)$ A	B	C	D	E	A Solid Content (wt%)	B Down Force (psi)	C Back Pressure (psi)	D Platen Speed (rpm)	E Polish Time (sec)
1	1	1	1	1	1	5%	4	0	20	40
2	1	2	2	2	2	5%	5	1	30	45
3	1	3	3	3	3	5%	6	2	40	50
4	1	4	4	4	4	5%	7	3	50	55
5	1	5	5	5	5	5%	8	3.5	60	60
6	2	1	4	2	3	10%	4	3	30	50
7	2	2	5	3	4	10%	5	3.5	40	55
8	2	3	1	4	5	10%	6	0	50	60
9	2	4	2	5	1	10%	7	1	60	40
10	2	5	3	1	2	10%	8	2	20	45
11	3	1	2	3	5	15%	4	1	40	60
12	3	2	3	4	1	15%	5	2	50	40
13	3	3	4	5	2	15%	6	3	60	45
14	3	4	5	1	3	15%	7	3.5	20	50
15	3	5	1	2	4	15%	8	0	30	55
16	4	1	5	4	2	20%	4	3.5	50	45
17	4	2	1	5	3	20%	5	0	60	50
18	4	3	2	1	4	20%	6	1	20	55
19	4	4	3	2	5	20%	7	2	30	60
20	4	5	4	3	1	20%	8	3	40	40
21	5	1	3	5	4	25%	4	2	60	55
22	5	2	4	1	5	25%	5	3	20	60
23	5	3	5	2	1	25%	6	3.5	30	40
24	5	4	1	3	2	25%	7	0	40	45
25	5	5	2	4	3	25%	8	1	50	50

Table 2. Experimental results and S/N ratio

	$L_{25}(5^5)$					MRR (A/S)	(S/N)	σ_{t2}	(S/N)
	A	B	C	D	E				
1	1	1	1	1	1	9.14	19.22	137.540	17.23
2	1	2	2	2	2	14.5	23.23	109.591	19.20
3	1	3	3	3	3	16.34	24.26	99.567	20.04
4	1	4	4	4	4	20.31	26.15	128.673	17.81
5	1	5	5	5	5	20.55	26.26	98.534	20.13
6	2	1	4	2	3	16.09	24.13	51.999	25.68
7	2	2	5	3	4	23.45	27.4	152.866	16.31
8	2	3	1	4	5	28.92	29.22	298.643	10.50
9	2	4	2	5	1	34.64	30.79	88.683	21.04
10	2	5	3	1	2	22.51	27.05	112.108	19.01
11	3	1	2	3	5	19.44	25.77	92.238	20.70
12	3	2	3	4	1	27.68	28.84	114.126	18.85
13	3	3	4	5	2	36.89	31.34	114.363	18.83
14	3	4	5	1	3	22.11	26.89	55.401	25.13
15	3	5	1	2	4	31.84	30.06	110.232	19.15
16	4	1	5	4	2	20.44	26.21	100.058	19.99
17	4	2	1	5	3	29.22	29.31	169.638	15.41
18	4	3	2	1	4	18.37	25.28	167.419	15.52
19	4	4	3	2	5	28.72	29.16	114.870	18.80
20	4	5	4	3	1	39.28	31.88	69.075	23.21
21	5	1	3	5	4	23.25	27.33	113.494	18.90
22	5	2	4	1	5	17.1	24.66	46.615	26.63
23	5	3	5	2	1	25.74	28.21	52.903	25.53
24	5	4	1	3	2	34.59	30.78	139.578	17.10
25	5	5	2	4	3	44.11	32.89	206.076	13.72

Table 3. Table of response for MRR

Factor / Level	A Solid Content (wt%)	B Down Force (psi)	C Back Pressure (psi)	D Platen Speed (rpm)	E Polishing time (sec)
1	23.80363	24.53263	27.71867	24.62076	27.78998
2	27.71891	26.68942	27.59314	26.95966	27.72017
3	28.58194	27.66443	27.32932	28.02038	27.49836
4	28.37092	28.75611	27.63418	28.66399	27.24576
5	28.77449	29.60729	26.97457	28.9851	26.99562

Table 4. Table of response for σ_{t2}

Factor / Level	A Solid Content (wt%)	B Down Force (psi)	C Back Pressure (psi)	D Platen Speed (rpm)	E Polishing time (sec)
1	18.88245	20.50176	15.8791	20.7043	21.17418
2	18.50823	19.28189	18.03852	21.6729	18.82889
3	20.53432	18.08461	19.11871	19.4741	19.99524
4	18.58753	19.97647	22.43348	16.8772	17.54044
5	20.37666	19.04445	21.41936	20.6428	19.35043

Table 5. Table of variance analysis for MRR

Factor	Sum of square	D. O. F.	Mean of square	Degree of contribution
A	86.2601	4	21.565	37.5417
B	77.4767	4	19.3692	33.719
C	1.83596	4	0.45899	0.79904
D	62.0033	4	15.5008	26.9848
E	2.19542	4	0.54886	0.95548
Total	229.771	20		100

Table 6. Table of variance analysis for σ_{t2}

Factor	Sum of square	D. O. F.	Mean of square	Degree of contribution
A	19.8066	4	4.95164	6.93045
B	17.0717	4	4.26793	5.97351
C	138.034	4	34.5086	48.2991
D	74.4472	4	18.6118	26.0496
E	36.4308	4	9.10771	12.7474
Total	285.791	20		100

Table 7. Verification experiment for optimal setting

	Original setting	Optimal setting
Solid Content	15 %	20 %
Down Force	7 psi	8 psi
Back Pressure	3 psi	3.2 psi
Platen Speed	20 rpm	41 rpm
Carrier Speed	25 rpm	25 rpm
Oscillation Speed	2 mm/sec	2 mm/sec
Polishing Time	60s	42s
Average of removal amount	1599.2 A°	1897.9 A°
Average of thickness after polishing	3509.9 A°	3149.9 A°
WIWNU of thickness after polishing	112.2 A°	118.0 A°
Experimental MRR	26.65 A°/s	45.19 A°/s
Experimental WIWNU	3.1963 %	3.7506 %